MW00581819

Promoting Research Integrity in a Global Environment

Promoting Research Integrity in a Global Environment

Editors

Tony Mayer
Nanyang Technological University, Singapore

Nicholas Steneck
University of Michigan, USA

NEW JERSEY • LONDON • SINGAPORE • BEIJING • SHANGHAI • HONG KONG • TAIPEI • CHENNAI

Published by

World Scientific Publishing Co. Pte. Ltd.

5 Toh Tuck Link, Singapore 596224

USA office: 27 Warren Street, Suite 401-402, Hackensack, NJ 07601

UK office: 57 Shelton Street, Covent Garden, London WC2H 9HE

British Library Cataloguing-in-Publication Data
A catalogue record for this book is available from the British Library.

PROMOTING RESEARCH INTEGRITY IN A GLOBAL ENVIRONMENT

ISBN-13 978-981-4340-97-7
ISBN-10 981-4340-97-9

In-house Editor: Ms. Sandhya Venkatesh

Typeset by Stallion Press
Email: enquiries@stallionpress.com

Printed in Singapore by World Scientific Printers.

PREFACE

The World Conferences on Research Integrity began in early 2005 as a modest effort to expand a US Office of Research Integrity (ORI) outreach programme to Europe. At the time, ORI provided small grants to support conferences as a way of promoting discussion on the responsible conduct of research. In early 2005, then Director of ORI, Chris Pascal, and Program Director, Larry Rhoades, agreed to provide a few days of travel support for a consultant working in their office, Nick Steneck, to identify a partner for a joint US-European research integrity conference. Based on a suggestion from Pieter Drenth, President of ALLEA, a meeting was set up with the European Science Foundation Chief Executive, Bertil Andersson. Out of that meeting emerged the collaboration that led to the initial World Conference in Lisbon (2007), the 2nd World Conference in Singapore (2010), the Singapore Statement on Research Integrity and the ongoing World Conferences on Research Integrity. Tony Mayer joined the effort shortly after the ESF meeting as ESF's representative and co-chair, with Nick Steneck, as the initiative rapidly took on a global perspective.

Planning for the inaugural World Conference on Research Integrity received its first major support from the European Commission in late 2005. Soon thereafter, support from the Portuguese European Union Presidency, under whose auspices the Conference was held, and the Gulbenkian Foundation, set the location and date for this first world event to gather and discuss the many issues relating to promoting integrity in research. Other major partners then joined over time, including the International Council of Science (ICSU), the Committee on Publication Ethics (COPE), the European Molecular Biological Organisation (EMBO) and the Japan Society for the Promotion of Science (JSPS), all of which were at the forefront of support for the Second World Conference and whose commitment to promoting research integrity has been greatly appreciated. The First Conference was an initial attempt at providing World awareness of research integrity and discussing strategies for harmonising policies and fostering responsible conduct in research. At the same time as the planning for the Conference commenced, the Organisation for Economic Co-operation and Development

(OECD) Global Science Forum, under the leadership of Canada and Japan, embarked on its own study on research integrity. This effort formed an important thread within and was reported at the First Conference.

Delegates at the Lisbon meeting strongly recommended that a second World Conference should be planned to address institutional leadership as one of its themes. It was further suggested that the Second Conference should be held in Asia to show that research integrity was a worldwide issue and not just confined to Europe and North America. Singapore took up the challenge and offered to host the Second World Conference in July 2010. Singapore's three major universities — Nanyang Technology University (NTU), National University of Singapore (NUS), and Singapore Management University (SMU) — the Agency for Science, Technology and Research (A*STAR), and the Singaporean Ministry of Education provided very generous support, with NTU acting as the lead organisational institution. ESF and ORI also continued their support along with the other organisations listed above. Mention also needs to be made of the many national agencies from around the World that supported the Conference through travel support to participants from their countries and to World Scientific Publishing Company for publishing these proceedings.

Looking back over the five years of work on the World Conferences, one factors stands out above all others as the key to success: commitment. Few if any in the world of research question the importance of integrity. It goes almost without saying that researchers "should take responsibility for the trustworthiness of their research." "Honesty" is imperative "in all aspects of research" (Singapore Statement). In principle, integrity in research is widely embraced. But doing something about research integrity is another matter. Leadership in this area requires a commitment to the promotion of integrity in research, even if there are no formal requirements or mandates to do so. ORI and ESF's initial commitment was essential to the launching of the World Conferences. Bertil Andersson's continued willingness to help has been indispensible throughout. Commitments by other leaders at crucial times allowed the initial US-European initiative to grow into a global venture.

Looking forward, steps are already underway to encourage use of the main product of the Second World Conference, the Singapore Statement (www.singporestatement.org), as a framework for developing, expanding, improving and harmonizing research integrity policies and other relevant guidelines or codes on a global basis. A small committee has also been established, under the leadership of Melissa Anderson, University of Minnesota, to plan a third World Conference on Research Integrity. While much has been accomplished, there is more work to do. Serious misconduct in research is still a problem throughout the world. Too many researchers cut corners and engage in questionable practices that unfortunately compromise the trustworthiness of the research record and the trust the public has in research. But most importantly, future *researchers* need to be better and more consistently made aware of the many responsibilities they have and *research institutions* need to work to provide working environments that encourage rather than discourage responsible behaviour. Over the past few decades, research has been globalized. We now need to complete the task of globalizing efforts to promote and monitor integrity in research.

Tony Mayer, Swindon UK
Nick Steneck, Ann Arbor, Michigan USA
March 2011

ACKNOWLEDGEMENTS

We wish to acknowledge the generous support for the Second World Conference that we received from the Singapore Ministry of Education, the Agency for Science, Technology and Research (A*STAR), Nanyang Technological University (NTU), which took the organizational responsibility, the National University of Singapore (NUS), the Singapore Management University (SMU) and the Singapore Tourism Board (Singapore Exhibition and Convention Bureau). We are also grateful for the ongoing commitment and support for both the First and the Second World Conferences from U.S. Office of Research Integrity (ORI), the European Science Foundation (ESF), the Japan Society for the Promotion of Science (JSPS), the International Council of Science (ICSU), the European Molecular Biology Organisation (EMBO), and the Committee on Publication Ethics (COPE) as well as the U.S. National Science Foundation (NSF) and the American Association for the Advancement of Science (AAAS). In addition, the Second Conference was able to attract participation from many countries around the World through the support of travel and other costs of national funding organisations (Australian Research Managers Society — ARMS; the Centre for Good Research Practice of Korea; China Association for Science and Technology; European Cooperation in Science and Technology — COST; European Federation of Good Clinical Practice; King Abdullah University of Science and Technology — Saudi Arabia; the National Science Council of Taiwan; the Research Councils UK; and Thomson Reuters). We are especially grateful to World Scientific for undertaking the publication of these Proceedings. Our especially thanks go to Professor Bertil Andersson for his personal commitment to both World Conferences, to Mme Jan Ho who provided enthusiastic and efficient administrative support for the Conference, to Mme Sandhya from World Scientific for her editorial skills and for getting us to adhere to schedules and to Professor Melissa Anderson for her contribution to the drafting of the Singapore Statement and for taking on responsibility as the organizer of the Third World Conference.

Tony Mayer and Nicholas Steneck

CONTENTS

SECTION I

WELCOMES

INTRODUCTION

The World Conferences on Research Integrity represent an effort to provide guidance for promoting integrity in research throughout the world. The First World Conference represented an initial discussion of problems and solutions, with the goal of assembling a community of researchers and administrators who could take key issues back to their countries and institutions for further discussion (http://www.esf.org/index.php?id=4479). The Second World Conference focused more specifically on leadership, challenging those who presented to identify issues or problems and suggest what was being done or could be done to address them (http://www.wcri2010.org/index.asp).

In all, 340 individuals from 51 countries attended the conference. Attendees included researchers, funders, representatives of research institutions (universities and research institutes) and research publishers. This opening section includes talks by five research leaders in Singapore. As Minister Ng Eng Hen explains, as a recent entrant into world-class research, Singapore has a special interest in building "a reputation of being able to produce high quality scientific outputs under strict ethical standards and academic integrity." Singapore's support for and participation in the Second World Conference represents its strong commitment to leadership in this area, as elaborated by the five welcome talks in this section.

The remainder for this volume presents papers and reports from the five key topics addressed both during the Conference and the Workshops held the day after the Conference:

Section 2: Research Integrity Structures,
Section 3: Research Misconduct,
Section 4: Codes of Conduct,
Section 5: Foster Responsible Research,
Section 6: Integrity Issues for Authors and Editors.

The final section, Section 7, stems from a special session on Research Integrity in the News, during with which climate change, dual-use research, and plagiarism detection were discussed.

OPENING ADDRESS BY THE MINISTER FOR EDUCATION AND SECOND MINISTER FOR DEFENCE

Ng Eng Hen

I am pleased to join you here today for the opening of the Second World Conference on Research Integrity. I am also glad that NTU, NUS and SMU are hosting this Conference.

In relative terms, Singapore entered the R&D arena only recently compared with other developed countries with established centres and on-going investments in diverse scientific fields. To gain credibility and be among the leaders in R&D globally, Singapore must build a reputation of being able to produce high quality scientific outputs under strict ethical standards and academic integrity.

1 SCIENCE AND RESEARCH AS A BUTTRESS FOR ECONOMIC GROWTH

The journey for Singapore to venture into R&D was not an intuitive one. Our forays into this field were as much a response to competition as it was a default choice when looking at limited options for our economic strategy. Fortuitously, we had started well by ensuring that our students had good capabilities in science and technology. This was a necessity because bereft of natural resources apart from human capital and without a hinterland, our economic strategies necessarily focused on providing highly skilled labour selling products to global markets. As we could not compete on size, we needed to leverage on science and technology to amplify our strengths and extend our reach.

This explains why barely two years after Singapore gained independence, we created a Science Council to advise the Government on matters relating to science and technology. But it would take longer, before the necessary conditions were in place for impactful R&D to be done here.

It was later in 1991, that the Government launched the first National Science and Technology 5-year Plan, which established the National Science and Technology Board, the successor to the Science Council. This Board was later reorganised to become the Agency for Science and Technology, or A*STAR, in 2001.

Part of this evolution to an economy centre on creating value through innovation and new ideas, was and is still driven by our neighbouring countries with cheaper factors of production, like land, labour and energy. Will Singapore succeed in this next lap where more is required of our human capital and with our competitive edge in traditional industries being eroded? We certainly hope so, because much of our economic progress depends on it.

But we have to be proactive and there are encouraging signs that we have at least, strong fundamentals in place. First, we have achieved high standards in maths and science among students, as shown in international comparisons such as the Trends in International Mathematics and Science Study or TIMSS. Their most recent report ranked Singapore students amongst the top 3 in the world for both maths and science achievement.

All of our Grade 10 students offer Mathematics and Science. More than 97% of our Grade 12 students offer Mathematics and 86% offer at least one Science subject. In Great Britain, this percentage drops to less than 15% for science and in America, the percentage of students offering Advanced Placement in science subjects is as low. We are fortunate to have a wide and high base to draw from and which must be continually cultivated to maintain that interest and edge for Singapore.

MOE will continue to put in resources to enhance the interest in and the learning of STEM — science, technology, engineering and maths at all levels — beginning from primary schools right through to our vocational institutes, polytechnics and universities. All our schools are equipped with science labs and resources. Our teachers are trained to elicit feedback and critique their delivery of content, to ensure that concepts are understood and better still visualised.

Part of the strategy to embed STEMs in our culture is to stretch our ablest. We encourage and facilitate for our top students to take part in the various International Science and Maths Olympiads and participate in overseas events such as the Intel International Science and Engineering Fair. The NUS Maths and Science High School is a specialised school for such talents.

These various initiatives have persuaded more than half of our students to pursue a science and technology degree at our universities. This is well above the average of 25% in OECD countries.

But these efforts will need to be maintained as the natural inclination for developed countries is a decline in the interest of STEM subjects. With economic prosperity and political stability, the motivation for students to study STEM subjects diminishes. Gradually, students will turn to study other subjects which they find intrinsically motivating and would lead to careers in finance and commercial sectors which offer higher monetary rewards. At the same time, we have to evolve our education system to facilitate creative thinking, refine thought processes and sharpen communication skills.

But to build expert mountain climbers in R&D, Singapore will also have to build not more petrochemical islands, airports or seaports, as we did in the past when we needed them, but R&D peaks. Over the last four years, five Research Centres of Excellence have been established within NUS and NTU, with topics ranging from Quantum Physics to Mechanobiology.

We are thankful that we have been able to attract leading experts in these fields of study, who will help us develop a broad and deep spread of scientific talent here. To name a few, we have Dr Edison Liu, Executive Director of the Genome Institute of Singapore and ex-President of the Human Genome Organisation who is a top cancer researcher. We have Prof Artur Ekert, a leading expert of quantum computing and cryptography now at NUS. We also have Prof Staffan Kjelleberg, who is internationally recognised for his studies in biofilms, starting up the Singapore Centre on Environmental Life Sciences Engineering at NTU.

In this quest to scale new heights at rarefied atmospheres, joint efforts and partnerships are crucial. We would be short-sighted and plain silly to try to do this alone. Locally, our universities, A*STAR and industry need to form strong bonds to form a triumvirate. A*STAR supports a wide range of industry-oriented research through its 14 research institutes and extramural research at our universities and hospital research centres. Collectively, the talent pool consists of bright graduates working side-by-side with leading scientists from around the world.

This wider economic strategy aimed at transforming Singapore into a knowledge-driven economy is driven at the highest levels of leadership, chaired by the Prime Minister, through the Research, Innovation and Enterprise Council or RIEC.

Needless to say, R&D is an expensive investment with long timelines. In 1991, Singapore's Gross Expenditure in R&D, or GERD, was just half a billion, less than 1% of Singapore's GDP then. We had only 28 research scientists and engineers per 10,000 labour force. Today, our GERD at $7 billion dollars annually is almost 3% of GDP. Nearly 1% of Singapore's workforce, or 26,000, are research scientists and engineers.

2 SINGAPORE'S CONTINUED COMMITMENT TO RESEARCH INVESTMENT

As with most long-term ventures, early harvests sustain both morale and commitment, and thankfully we have a few. Take electronics for example. We are one of the world's leading manufacturing sites for research tools and diagnostics instruments, supplying more than half of the world's micro-arrays and the global demand for thermal cyclers. Singapore also accounts for 40% of the world's HD media volume. This advantage can be sustained by the presence of corporate labs from electronics powerhouses like Seagate and Showa Denko.

Clean technology, or cleantech, is another example. Many global players have recognised the research potential at our universities, and have set up corporate labs at their campuses. GE Water set up a $130 million dollars R&D centre in NUS, while Bosch is investing $30 million dollars in a lab looking at organo-photovoltaics in NTU. These have also drawn in leading wind and solar companies like Vestas and REC to set up a manufacturing presence in Singapore. Altogether, the cleantech industry in Singapore is expected to generate $3.4 billion dollars a year in value-add and employ 18,000 people by 2015.

Singapore will also soon build up a 50-hectare CleanTech Park, our first eco-business park, immediately adjacent to NTU. This is part of the $1 billion dollars Singapore Sustainable Blueprint announced last year that will hopefully put Singapore on the world map as a centre for developing, testing and commercialising green technology.

Singapore is also making head-way in the field of Bio-Medical Science. A*STAR's Institute of Materials Research and Engineering, for example, is working with Advanced Technologies & Regenerative Medicine through its Singapore affiliate Johnson & Johnson to develop customised artificial "nano-skin" for implantable medical devices to make them more biocompatible. Researchers at the Cancer Science Institute, another Research Centre of Excellence in NUS, have recently reported a novel combination of drugs that can potentially halve therapy costs and decrease the side effects of treatment for advanced breast cancer patients. We hope for more such innovative scientific breakthroughs from our universities and research institutes working with multinational giants like Roche, Novartis and GlaxoSmithKline here.

3 RESEARCH INTEGRITY AS PUBLIC TRUST

I mentioned earlier that Singapore was a relative newcomer in this high-stakes game. We should note that competitors, with much larger economies, are not standing still. The Obama Administration has pledged to devote more than 3% of America's GDP to R&D. South Korea is aiming for 5%. France is spending €4.4 billion euros on a new "super university" to rival the top universities of today. But this race is virtuous and Singapore will form more linkages for international collaboration and cooperation.

High standards of integrity and ethics in research are crucial in Singapore's quest to be a R&D node in this global network. Knowledge without integrity can harm. The 2nd World Conference on Research Integrity is therefore timely and will assist us in reviewing our systems and procedures for our universities and research institutes.

4 CONFERENCE OUTCOMES

To this end, I am glad that this conference has already set for itself a discrete deliverable. It aims to crystallise one of the main recommendations from the First World Conference on Research Integrity — the need for consistent institutional and national policies. It will work towards developing a set of fundamental and basic principles to be agreed by consensus at this meeting, and which will be known collectively as the 'Singapore Statement'. It is my hope that this Statement will serve as a basic document for a global code of conduct and protocols, that can be adapted and used by individual countries for their own institutions to address the issues pertaining to research integrity and good research practices.

5 CONCLUSION

I wish to thank all the sponsors of this conference, and NTU, NUS, SMU and A*STAR. And also the US Office of Research Integrity and the European Science Foundation, which initiated the inaugural World Conference on Research Integrity in 2007. I want to thank all the other major contributors who have provided support for this meeting in one way or another.

I wish you all exciting, engaging and enlightening discussions during the next few days. To our guests from all over the world, I wish you a most pleasant stay in Singapore.

It gives me great pleasure to declare the 2nd World Conference on Research Integrity officially open.

Thank you.

WELCOME BY THE PRESIDENT OF NANYANG TECHNOLOGICAL UNIVERSITY

Su Guaning

To our eminent speakers and delegates who have come from all over the world, I bid you a very warm welcome to Singapore. We are indeed honoured to have you here with us. We have over 300 participants from over 50 countries gathered here today, making our conference a truly international one.

1 BACKGROUND

The First World Conference on Research Integrity was organised by the United States Office of Research Integrity (ORI) and the European Science Foundation (ESF) in Lisbon, Portugal, in September 2007. It was an opportunity for us to discuss key points of research misconduct policies and to strategise how we should inculcate responsible conduct in research. Specifically, the conference addressed research integrity issues — defined mainly as falsification, fabrication and plagiarism.

Following the success of that inaugural conference, it was agreed that we should meet again at a second conference, held preferably in Asia. Singapore, through its three universities — Nanyang Technological University (NTU), National University of Singapore (NUS) and Singapore Management University (SMU) — and the national research agency, Agency for Science, Technology And Research (A*STAR) offered to host the meeting. And here we are now in Singapore to continue the discussions and build on the foundations started three years ago in Lisbon.

Singapore is honoured to host this highly regarded event, supported not only by the three Singaporean universities and A*STAR, but also by the Ministry of Education, which shows the emphasis that Singapore places on educating its students on the importance of research integrity.

NTU is proud to be appointed the lead organiser and host for this event, bringing together renowned experts in their respective fields to address the vital issue of research integrity. Cutting across all disciplines, research integrity has become increasingly important today, given that innovation and R&D are key drivers of economic growth worldwide.

2 CONFERENCE THEME

The theme of this year's conference is **'Leadership Challenges and Responses'**. With many research activities now taking on a global dimension, it is imperative to discuss positive approaches towards inculcating best research integrity practices, including examining the role of academic publications in setting the standards for integrity.

We have with us today representatives from research funding organisations, universities, as well as other research agencies. I hope that this four-day conference — a platform for thought-leaders, academics and researchers to share their ideas and views on common research integrity issues — will challenge all delegates, in particular the heads of institutions, to think more about the leadership challenges and responses in research integrity which may in turn inspire new and practicable standards in the field.

3 CONFERENCE OUTCOMES

This conference, however, will not be just another "talk shop". By the end of the event, we intend to formulate a set of recommendations, to be known as the 'Singapore Statement'. This statement, to be crafted based on the discussions during the conference, will focus on recommendations on four key aspects of research integrity, namely:

1. A national and international framework for promoting research integrity and responding to misconduct;
2. Global codes of conduct and best practices for research integrity;
3. Coordinated principles and strategies for training students and researchers in best practices, and
4. Best practices for academic editors and publishers.

The Singapore Statement will be a benchmark for the future. All conference delegates are encouraged to get involved in its development through an online discussion forum. The debate has started even as I speak. This exchange of views and information will continue over the entire duration of the conference.

In an increasingly globalised world, there is a critical need to develop guidelines and recommendations for promoting integrity in research at an international level. To this end, the Singapore Statement shall serve as a fundamental and landmark document containing basic principles that can be used as a standard for research integrity throughout the world.

4 RESEARCH INTEGRITY AND EDUCATION

Such a standard will be especially important for educators and educational institutions like NTU, growing rapidly as major research universities. It will serve as a guideline against which our behaviour can be assessed, and be a point of reference to deal with infractions in research integrity.

Indeed, it is important to imbibe good values and practices early on in our young researchers, starting at the undergraduate level. Not only should we guide them on what is inappropriate behaviour, but more positively, we must educate them about best practice in research.

At NTU, we have taken steps to introduce a clear policy, based of course, with due acknowledgement, on the model provided by the US Office of Research Integrity. We have a zero tolerance policy for anyone, whatever their status, who breaches the university's research integrity norms. In addition, we are developing, in collaboration with a number of universities around the world, programmes to educate our young researchers, as part of a consortium led by Epigeum from London.

Another step we have taken is to introduce a clause in our international partnership agreements pertaining to integrity. The clause commits our partners and NTU to adopting best practices and undertaking joint action in handling any cases of breach of integrity which may arise.

But beyond educators and university leaders, it is equally important that the leaders of funding agencies as well as research institutions are committed to the best practices in research. Indeed, I am very pleased and heartened to see many delegates holding leadership positions attending this conference. This bodes well for the future of research integrity worldwide.

5 RECOGNITION OF PLANNING COMMITTEE MEMBERS AND SPONSORS

At this juncture, I would like to take the opportunity to thank the Local Organising Committee, chaired by Prof Tjin Swee Chuan, and the International Planning Committee, headed by Prof Nicholas Steneck and Mr Tony

Mayer, for their hard work and effort in planning and coordinating this event. Prof Steneck, who is widely acknowledged as a world expert in this area, has been instrumental in putting together the conference programme.

I would also like to thank the Ministry of Education, A*STAR, as well as our three Singaporean universities for their support in making this conference possible.

Other organisations that have contributed to this event include the European Science Foundation, the US Office of Research Integrity, the Committee on Publication Ethics, the US National Science Foundation, and the European Molecular Biology Organisation.

In addition, we received support from the Japan Society for the Promotion of Science, International Council for Science, European Forum for Good Clinical Practice, China Association for Science and Technology, American Association for the Advancement of Science, European Cooperation in Science and Technology, Research Councils UK, the Korean Centre for Good Research Practice, South Africa's National Research Foundation, the National Science Council, Taiwan, King Abdullah University of Science & Technology and Thomson Reuters.

Last but not least, I would also like to acknowledge the Singapore Tourism Board for their supporting grant for this event.

In closing, I encourage delegates to participate actively in the interesting discussions over the next four days. I wish everyone a successful and fruitful conference.

Thank you.

WELCOME BY THE CHAIRMAN OF A*STAR

Lim Chuan Poh

It gives me great pleasure to add my warm welcome to all of you to the 2nd World Conference on Research Integrity in Singapore.

A*STAR considers research integrity as a key and fundamental pillar of any organization that is serious about the Science and Technology (S&T) enterprise. For this reason, A*STAR is proud to be one of the co-sponsors for this conference.

I am heartened to learn that more than 350 concerned scientists, scientific leaders and magazine editors from Singapore and around the world share a similar view and have registered to attend this conference.

This augurs well for what we are trying to achieve in this conference.

1 A*STAR's PRIORITY ON RESEARCH INTEGRITY

Across the world, and in particular in Asia, the scientific enterprise has grown significantly over the last ten years. In the process, the competition to be the first to publish has also phenomenally intensified. Under such intense competition, lapses in research conduct can and do happen. Each time it happens, it has a huge impact not just for the individuals involved, but the institution and sometimes even the host countries of the research efforts. It can even potentially derail international efforts in combating global challenges. All of us in this room and many outside, therefore, share a common desire to shape and create a research environment that will be more conducive for and better promote responsible conduct of research (RCR). A*STAR regards research integrity as a critical pillar of our enterprise for four main reasons.

Firstly, our research is becoming more collaborative with many partnerships transcending both organizational as well as national borders. We are involved in both national scale research consortia as well as many bilateral and multilateral research partnerships with counterpart agencies in other countries. We also witnessed many more research collaborations with private sector enterprises. For such cross-border and multi-organisational collaborations to be effective and successful, there must be mutual confidence and trust in what each party is bringing to the partnerships or collaborations and how each party goes about conducting its business. A strong reputation of upholding a generally accepted common code of research ethics and integrity will go a long way to facilitate such complex collaborations and partnerships.

Secondly, the notion of 'publish or perish' is a very real fear that exists amongst many budding as well as established researchers. Given the increased research efforts across the world, this fear is only going to increase. As a research organization, A*STAR strives to create an environment that promotes and nurtures the value of responsible conduct in research (RCR). This, of course, is easier said than done and requires exemplary conduct on a day to day basis by the senior researchers and scientific leadership. The junior scientists generally take their cue from their seniors.

Thirdly, through our Biomedical Research Council, A*STAR is one of the key partners of the academic medical centres and hospitals in translational and clinical research. In this regard, the conduct of clinical trials involving human subjects and laboratory experiments with animals have to be subject to strict bioethical codes and requirements. And finally, for an organization that depends largely on public monies for research, A*STAR, and in particular our scientists, must continue to win the confidence of the public in how we conduct our research. This is what A*STAR needs to do to remain credible as an organization and to sustain strong public support for our endeavours.

Having a framework for research ethics and integrity is thus of utmost importance to A*STAR and the wider R&D community.

2 CHALLENGES RELATED TO RESEARCH INTEGRITY

However, it is one thing to deal with the issue of research integrity within an organization or a country like Singapore. It is quite a different matter to put in place a common code globally as we do have significant differences in our regulations and codes. For instance, I was told that there is no common definition world-wide for research misconduct, conflict of interest or plagiarism. Even where there is general agreement on key elements of research behaviour, such as the need to restrict authorship to individuals who make substantive contributions to the research or to provide protection for research subjects, the policies that implement this agreement can vary widely from country to country and organization to organization. The research community worldwide therefore needs to address these problems and to establish clear best practice frameworks at an international level.

We should also consider some of the greatest challenges that mankind currently face — issues such as climate change, environmental degradation, infectious diseases, and depletion of energy resources that have significant impact globally. Given the scale of the challenge, the only sensible way for us to go forward is to call on the multitude of expertise to work collaboratively across the world. This ability to work together will be greatly enhanced by a generally accepted code of research ethics and conduct. Any seeming lack of research integrity here can significantly undermine public confidence in the credibility of the efforts and therefore the influence and impact of the research outcomes.

3 CONCLUSION

On this note, I hope that this conference will provide a productive opportunity for the exchange of ideas and sharing of experiences on research integrity. I wish everyone fruitful discussions over the next two days. I also hope that our overseas delegates will bring back fond memories of their time in Singapore.

Thank you very much.

WELCOME BY THE VICE PRESIDENT FOR RESEARCH STRATEGY, NATIONAL UNIVERSITY OF SINGAPORE

Seeram Ramakrishna

On behalf of the National University of Singapore, I join Dr Su Guaning, President of Nanyang Technological University, to welcome all of you to the 2nd World Conference on Research Integrity.

We are witnessing transformation of global research and innovation enterprise with growing R&D expenditure and innovation activities in Asia. The world is now spending more than a trillion dollars on research and development, perhaps the highest in modern history. Nearly one-third of this sum comes from public funds. Naturally there are more expectations on the research community to deliver scientific breakthroughs of grand challenges of the day.

As researchers focus on making scientific progress and impact, the emphasis on research integrity cannot be understated. This is more so in a globalised world of community of researchers with diversity in languages, culture and practices. Sustained efforts with due attention must be made by various stakeholders and researchers to nurture a culture of Research Integrity and higher standards.

NUS has an international community of faculty and students from over 100 countries. Recognising the importance and to promote a culture of Research Integrity, NUS took the lead to write the Research Integrity code in 2006. We called it *NUS Code & Procedures on Research Integrity.* Since then, we have continued to make regular revisions to keep it updated and relevant to the evolving research landscape. Research Integrity is all about Code of Honour, confidence and TRUST. In a recent book by Stephen M.R. Covey, *Speed of Trust,* it said that with high trust level, things get done faster and at lower cost, whereas low trust leads to things being done slower and at higher cost. This is intuitive and yet profound.

In closing, I wish all of you a fruitful and successful conference. For our overseas guests, we invite you to visit some of our scenic areas and enjoy vast variety of Asian and international cuisines.

Thank you.

WELCOME BY THE PRESIDENT OF SINGAPORE MANAGEMENT UNIVERSITY

Howard Hunter

Dr Ng Eng Hen, Minister for Education and Second Minister for Defence

My colleagues, friends, guests of Singapore,

Welcome to Singapore on this beautiful but slightly damp morning. I know this is going to be an interesting conference and I am going to take the liberty of setting some of my prepared remarks to one side because the prepared remarks are essentially the same as those already presented by my friends and colleagues Professors Tan Chorh Chuan and Su Guaning. That should not be surprising because we have been engaged in the same kind of endeavors for some period of time. I simply want to share with you some thoughts from 34 years in higher education, 22 of them in senior administrative positions, as a dean or a provost or a president. For a number of those years I was at Emory University in Atlanta where I had to deal with, on a quite regular basis, a number of the research integrity issues which you will be talking about over the next few days.

As Lim Chuan Poh of A*STAR was talking I remembered, for instance, serving on the university committee that looked after the welfare of laboratory research animals, With a large primate centre and another 80 to 100 thousand rats and other animals within the University, animal care was s a big issue. One thing that I distinctly remember is that the air quality standards for rats in the psychology department were much higher than the air quality standards set for the faculty and the staff. I always found that fact to be helpful in keeping things in perspective.

We all agree on certain core principles. The discovery of knowledge is one of the central missions of universities and of research institutions. The pursuit of knowledge satisfies the innate drive of the human curiosity. The discovery of new knowledge can help make the world a better, safer, more productive place. And we can all agree that this research should be and must be done with integrity and the results made available to the larger world to encourage further study and understanding. It is simple to state the core principles, and no one is likely to disagree. Implementation is much more difficult.

Consider first our students. We start them out, whether in SMU, NUS, NTU or any other major university in the world, with clear statements of principle. For example, we tell them not to plagiarize. Do not use the work of others as your own without permission and attribution. Doing so is theft. The scientific method requires interested, honest, rigorous and sceptical inquiry. We all know that today, a huge amount of information is freely available, and widely distributed, often without any filtration or peer review. I share with you just one recent and amusing example that has to do with rain here in Singapore. Last month, there was a minor flood on Orchard Road which was a mess for a little while and upsetting to shop owners in the area. A couple of cheeky students at SMU — most students at SMU are kind of cheeky — photoshopped a photograph of our dragon boat team from Kallang Basin onto flooded Orchard Road and sent it out electronically to friends around the world. I know it went around the world because I received at least 25 emails from friends in the US saying "My goodness, what is happening on Orchard Road?" No one ever seemed to ask the question "is this photo a real one?" And that's just a minor, amusing and harmless example of the problems we face all the time, with the free and wide distribution of information. We want to encourage wide and open distribution, but, at the same time, have some means to test for accuracy.

Good research, particularly in biomedical science and other areas, is often extremely expensive. Relationships with funders, whether they be government or private, can lead to a host of questions about disclosure and integrity. A former colleague at Emory has been much in the news. I will not mention his name, but he was a very distinguished researcher in the field of psychiatry who had a major laboratory, who published large numbers of papers, many of which received favourable coverage. He failed to disclose properly that he was receiving substantial consulting fees from certain large pharmaceutical companies that were interested in the research. This was not direct funding of the research but rather separate consulting fees in the seven-figure range. And as a result, not only did he lose his position at the university, but he compromised the integrity of all the work in which many people had been involved for a long period of time. His non-disclosure affected the reputations of the people who worked in his lab and his other colleagues. The ripple effects were tremendous. Many agree that the fundamental research itself was sound, but it will have to be subjected to replication studies for verification. There is also a tendency to move towards research that focuses more on immediate commercialization and return on investment, than deeper or purely curiosity driven science which is sometimes serendipitous. Such a focus can lead to more temptations for those who are not as careful as they should be about disclosure and research integrity.

In our university, the focus is on the social sciences and management disciplines together with law, accountancy, and computer science. There is less laboratory work and more field research. Much of the field research involves human subject surveys or observations. The demand for research integrity is no less than in the biomedical sciences or the physical sciences. We established in the earlier days of the University an Institutional Review Board which is organized according to the highest international standards. To set an example for all of our faculty and students, we require not only that the researchers comply with IRB standards but anything we do internally which involves surveys or any kind of human subject interviews, must be cleared by the IRB whether it is a survey by the Human Resources Department or something from the Finance Department. This process has turned out to be a good teaching tool for our students. Last year, for instance, our Student Association wanted to do an engagement survey of students. Before they could do so, they had to clear the survey and methods with faculty who teach social science methodology and then they had to make the presentation to the IRB and get full IRB approval before it could go forward. The result was a well-done survey which was useful to the Student Association and to the University leadership. It was also an excellent pedagogical experience for the students, and for the faculty and staff who worked with them. Embedding the principles of careful, honest research into the educational experience itself, I think, is very important.

I close with that by saying I join my colleagues in welcoming you here. I am very glad this conference is in Singapore because some of the most interesting and dynamic research work in the world is happening on this island and in other parts of Asia. The issues before this conference are very much at the forefront of what we are trying to do in our universities every day. Your presence in Singapore highlights the importance of research integrity to the mission of the universities and our research institutes. I welcome you and I hope you enjoy your stay in Singapore and that you have a fruitful time together.

SECTION II

RESEARCH INTEGRITY STRUCTURES

INTRODUCTION

Promoting integrity and responding to integrity begins with structures — international, national, institutional and local. The chapters in this Section present examples of structures from across the World for handling misconduct cases and for the promotion of good research practice. It is naturally very heterogeneous, reflecting the diversity of national and local organisational structures of research, as well as local conditions, traditions and history.

Boesz opened this discussion of challenges in this area by highlighting the problems caused by policy diversity in a rapidly globalising research world. This starts from the problem of creating harmonised systems even within one country, where there may be complex divisions of responsibility at the local level. She goes on to describe the attempts that have been made to arrive at some common guidance through the Organisation for Economic Co-operation and Development, Global Science Forum (OECD–GSF). Boesz highlights both the need for an agreed framework coupled with resources necessary to conduct international investigations. The OECD–GSF recommended adoption of 'Boilerplate' text that could be incorporated into international research agreements. Whatever form of words is used — and the 'Boilerplate' is a compromise — the key message is that a clause covering cooperation to promote research integrity and a commitment to jointly investigate alleged cases of misconduct should be the norm in all such international agreements at whatever level. Boesz points out that an underlying base condition, as recommended in the ESF report and further recommended in the Singapore Statement, is that each country has an obligation to develop its own structure on which international action can be based.

Alper reinforces this message by stressing the role of institutional leadership in addressing research integrity. He sets a high bar for such an achievement. In addition, he advocates that learned and professional societies have a significant role to play in such leadership. Overall, academic and research leadership is necessary to promote education in research integrity as the most positive way forward. Setting the bar high, promoting education and dealing sternly with miscreants is necessary to improve performance and develop and maintain public trust in research.

Lee Eng Hin describes the development of research in the Conference host country — Singapore. Singapore is a rapidly emerging research nation, which couples a policy of putting research at the heart of economic development with its ethos and reputation for incorruptibility. Lee describes the developments within the national research agency, A*STAR, in its aims to foster integrity within both its intra- and extra-mural research. This is a challenge faced by many such 'hybrid' agencies around the world.

Halliday outlines the developments taking place at a regional level within Europe and introduces the Code of Conduct developed by the European Science Foundation (ESF) and ALLEA as an example of what may be achieved when a positive spirit of cooperation exists between national agencies in several countries. In bringing agencies together from across Europe, ESF had also stimulated action by those countries currently lacking structures as well as encouraging those with working systems to review their effectiveness. Halliday, then discussed public trust in research, based on his experience in heading a major UK funding agency. He concludes with the suggestion that research has changed its character and so the methods of traditional science must change in step, especially in the new information age, with an emphasis on transparency and better record keeping.

Other contributors then describe the situations in a variety of countries and the steps being taken to build structures to promote research integrity and deal with misconduct. Heslegrave describes the complex processes and compromises necessary to establish a national system, as has been the recent experience in Canada. Rumball and O'Neill pose questions arising from New Zealand that demand the unambiguous commitment of institutional leadership. Bossi pays particular attention to scepticism and mistrust within society to academic self-policing — a theme that occured in many of the Conference contributions. He concludes that there has to be continuous action to promote research integrity, demonstrable to the public, in order to overcome this scepticism. Masui, in outlining the research integrity structures in Japan, with special emphasis on biomedical research, poses a behavioural question and one to which the community must give serious thought. He comes down on the side of both the freedom of researchers while reinforcing individual responsibility, which is, of course, reiterated in the Singapore Statement with its emphasis on individual responsibility. Finally, Ren describes the Australian experience in which action to promote research integrity is being taken at a variety of levels with institutional leadership being to the fore. He goes on to advocate the use of the Singapore Statement as the basis for incorporating research integrity into future international research collaboration.

Cooperation across borders is recommended by many speakers, but may be difficult due to language and other differences, even when there is a willingness to work together. This question is especially pertinent in Finland, a country with well-established guidelines and procedures, where the Finnish language does not have a word for 'integrity' and so 'ethics' is used not only for ethical matters but also as the translation for integrity. From another European perspective, de Hen describes how the European Network of Research Integrity Offices (ENRIO) has enabled countries and organisations to learn from each other — a regional example that could be widened to a global level, something that was considered by the OECD–GSF. Yudin describes the situation in Russia and the Commonwealth of Independent States (CIS) and draws attention to the suspicions of the researchers themselves at attempts to conduct surveys of research integrity behaviours based on their previous experiences under communism in the 20th century. This shows the barriers that frequently need to be overcome both within and outside the research community.

The overall message in these chapters is clear. We have to accommodate large variety both within nations and between them while at the same time striving for common ground, which can be provided by the Singapore Statement. It is also recognised that public trust in research has to be earned.

CHAPTER 1

DEVELOPING RESEARCH INTEGRITY STRUCTURES: NATIONALLY AND INTERNATIONALLY

Christine C Boesz

As research agendas continue to evolve, becoming more complex and involving more collaborators, the research support systems must also evolve to keep up with the increasing demands of modern research. An important, and some would argue a critical, aspect of the evolution is the development of structures that will support research[1] integrity across geopolitical boundaries and all cultures. Important compliance issues span a wide range of governmental, sponsor, and academic requirements, intended to protect the environment and individuals from possible harmful effects. At times these compliance initiatives place burdens on the research community, but have become essential safeguards to protect the public and its interest. Research misconduct is a reality although its prevalence on a global basis is unknown and is hard to estimate. There are, however, plenty of examples of poor behavior and lapses in judgment. The question is: How can integrity structures support the research yet handle difficulties as they arise? The task is difficult, and involves education, organizational responsibilities, and individual commitment. The following discussion will focus on one area of

[1]Andersen, M. and Steneck, N. (eds.) (2010). *International Research Collaborations. Much to be Gained, Many Ways to Get in Trouble.* NY and London: Routledge. This text describes ways in which international collaborations can be structure to be productive and compliant. It also describes examines threats to the success of these collaborations, recognizing the importance of integrity structures to such the international research efforts.

risk and interest to the research community and its sponsors: how to handle allegations of research misconduct.

Such allegations can be very damaging to the research enterprise and the individuals involved. The topic is a non-trivial issue with some very difficult challenges. Although international collaborations bring a rich and rewarding dimension to the research enterprise, there are risks associated with such arrangements.[2] With the growth in international collaborations, a logical outgrowth is the development of international policies and guidelines to respond to such allegations. Today the challenges are to better understand how to harmonize the various policies and procedures that have sprung up around the globe and to better share resources and expertise when investigating allegations. Yet the process of building an integrity structure begins at home. Research institutions, whether public or private, that have good integrity structures, will have good outcomes, including the handling of allegations of research misconduct.

When an allegation of misconduct occurs, the daunting challenge is to handle it in a fair and timely manner, with integrity and accuracy. When the allegation is investigated, the results and consequences of any actions should be communicated to all stakeholders, with emphasis on lessons learned. Communication is important because it helps the research community to understand not only the failure, but also the consequence of such bad or inappropriate behavior. Too often the consequences of bad or poor research practices are simply not understood. In general, communication of lessons learned and consequences supports advances in integrity and fosters good research practices.

Today's challenges in handling research misconduct may be more complex because of the involvement of multiple researchers, in as many institutions, in more than one country, with varying degrees of political and social harmony. Also, varying scientific, social and political interests contribute to the challenges and may sometimes impede the investigators' progress, or even thwart the initiation of an investigation. To reduce the negative effect of these interests, it is important to have unambiguous principles of expected behavior, clear definitions of misconduct and explicit rules and procedures on how to handle and investigate and follow-up adjudications. Furthermore, the importance of having appropriate structures in place are fundamentally important to having integrity in the investigative process itself.

[2]Ibid, Boesz and Fischer (2011). p. 123.

Where does one look for guidance in developing a good integrity structure? The Global Science Forum (GSF) of the Office of Economic Cooperation and Development (OECD) crafted one response. The OECD is a 33-country treaty organization that provides a forum for member and observing nations to compare and develop a variety of best practices to assist each other in market development.[3] Its GSF focuses on issues that have multiple-nation interest in fundamental science research.[4]

The GSF was initially approached in 2006 by the government of Japan with concerns about the global impact on science of serious research misconduct. The GSF agreed to accept the challenge of exploring best practices and developing some guidance in what to do with matters of misconduct in science. An International Steering Committee was established to convene a workshop in Tokyo during 2007. This resulted with a report accepted by the GSF, Best Practices for ensuring Scientific Integrity and Preventing Misconduct.[5] The participants were representatives of 23 governments that take part in GSF issues. OECD staff and experts in research integrity supported the process. During the process of developing the concepts and recommendations discussed in the report, numerous discussions focused on the root causes of misconduct, options for dealing with allegations, formal processes for investigating serious case, and prevention activities. Consensus among participants was achieved although geopolitical, academic and cultural differences were represented. The findings and recommendations of the Report are intended to apply to all domains of basic and applied research, including physical and life sciences, social and behavioral sciences, and the humanities.[6] The highlights of this report are summarized below.

First, after considering a broad range of inappropriate and bad behaviors, core "misconduct in research" was defined as plagiarism, falsification, and fabrication of data.[7] Initially, a few argued that plagiarism was a

---

[3]Refer to www.oecd.org for background information.

[4]The author was the United States representative to the GSF for the work related to research integrity.

[5]OECD–GSF (2007). *Best Practices for Ensuring Scientific Integrity and Preventing Misconduct.* Refer to www.oecd.org/dataoecd/37/17/40188303.pdf

[6]The report referenced above explains this rationale.

[7]These definitions are the same as the definitions used in the Untied States, issued in a policy statement in 2000 by the White House's Office of Science and Technology Policy.

victimless transgression and, therefore, the consequences were not serious to the scientific enterprise. However, the majority believed that the use of another's ideas without attribution, whether specifically printed information or stolen as a concept, was a serious misconduct and struck at the heart of the research enterprise. Within countries that have small research communities, there is a heavy reliance on reviewers from other nations. There is a concern that the international composition of review teams leads to the pilfering of ideas in proposals reviewed. Since most funders of research, particularly governments, use some system of peer review to judge which proposals get funded, unchecked plagiarism can erode the integrity of a peer review process and contribute to corruption in the funding decisions. As the funders' interests were paramount among GSF members, the majority view became the consensus position. While there are no statistics on the prevalence of such behavior, the concern alone is sufficient to worry the conscientious persons responsible for peer review processes.

Data fabrication and falsification seem like simple concepts. However, complex manners by which data is collected and manipulated led to intense discussions over how much, if any, tolerance can be permitted. The response was clear. There should be no tolerance when it comes to making up data, selectively excluding it from analyses, misinterpreting it to obtain desired result, or misrepresenting images in publications.

The OECD report details a wide range of inappropriate behavior that may be called misconduct by researchers, including but not limited poor research practices, poor personal behavior, poor publication practices, and financial wrongdoing.[8] While these activities are unfortunate, and even deplorable, the GSF Committee decided that the research community itself would best consider the broader range of misconduct. Plagiarism in proposing, performing or reviewing research or fabrication and falsification of data and reporting it as research results amounts to scientific fraud. Thus, the notion of intent to mislead is linked to the core definitions. Such acts of misconduct are not accidental, but rather deliberate actions by the researcher to behave badly. Because such deliberate actions by the researcher can lead to harm to society and to individuals, these acts of misconduct are serious, and possibly even criminal. In reality, damage to the scientific record can be very costly to the research enterprise. Costly, not only in terms of time and

[8]The OECD 2007 Report contains a useful table that classifies six types of misconduct, a spectrum of behavior that is considered unacceptable.

resources, but also in terms of the public's trust. The integrity of research is linked to the public's willingness to support research with both money and belief in the public good that comes from sound and safe research. Hence, the importance of integrity structures in supporting the research enterprise.

The GSF Report identified options for dealing with misconduct allegations, identified steps of fair and timely investigations, and discussed causes, factors and prevention activities.[9] Because allegations often arise in the minds of students, colleagues, and other researchers, it is important for intuitions, whether government or private, to have policy and procedures to handle allegations, including mechanisms for receiving allegations. The GSF Report spells out a series of questions that are important to consider when developing process to deal with allegations. As in any investigative procedure, it is important for every step of the process to have legitimacy since the reputations of whistleblowers and researchers can be damaged and not be easily restored. Also, as previously mentioned, the public must have confidence in the process.

There are a few basic principles that should be followed. First, the process should be based on a legal framework that is clear as to what organization has the authority to handle an allegation and who within that organization has the responsibility to do so. The more explicitly stated the better to avoid legal challenges that may develop further into the process. Also, organizational placement of this authority within an institution is important. The key concept here is independence. Independence of the investigator and investigative team is important to assure fairness and to prevent impartiality. For example, it is not appropriate for the investigators to report to the research management under investigation, nor to have significant conflicts of interest. While the notion of independence may be intuitively obvious, achieving the desired level of independence in an operational setting is often difficult to achieve. The temptation to have control of investigations under control of only the research community should be avoided.

Secondly, the investigators need to have independent resources, both monetary and human expertise. Without an appropriate level of resources within the control of the investigator, the investigator may not be able

[9]The OECD Report 2007 makes clear that certain practical and administrative aspects of dealing with misconduct are responsibilities of government that cannot be delegated to the scientific community, e.g., accountability for public funds, public safety.

to conduct a fair and timely investigation. An investigative budget is a critical matter that needs to be worked out in advance. Administrative independence also is linked to organizational placement. The investigator must be able to pursue the facts without being hindered by bureaucratic barriers or fear of retribution. In short, an investigation is a fact-finding adventure that must be conducted with the highest level of accuracy and integrity. There can be no question of bias because of where the investigator sits within an organization or where he or she gets the needed resources.[10]

The second GSF report, Investigating Research Misconduct Allegations in International Collaborative Research Project, was developed as a practical guide.[11] As a follow-on to the first project, the United States representative suggested that differences within and between national policies create challenges for those charged with investigating allegations of misconduct within the context of international collaborations. Guidance in this area would prove useful given the increasing number of international collaborations.[12] A new GSF Committee was established to address questions such as what happens if two or more nations' policies are at odds with each other?[13] What country takes the lead in an investigation? How can participants from various countries assist each other?

One difference among countries seems to lie in the basic definitions of misconduct and how deeply they reach into the researchers' behaviors. Some countries go beyond the basic three: plagiarism, fabrication, and falsification of data. For example, the core definition may be broadened to include the notion of a significant departure from accepted practices of the research community. Such a broad definition may make it difficult for the investigator to identify agreed upon standards upon which to base an investigation. Tighter, concise and precise definitions make the standards clear. Also, the

[10]Within the United States' National Science Foundation, independence of the investigator is assured by federal statute if conducted by the Inspector General. However, when the case is referred to the academic institution where the accused is employed, the federal role is to review the institution's report for fairness. Issues of independence would arise during this review.

[11]OECD–GSF (2009). *Investigating Research Misconduct Allegations in International Collaborative Research Project: A practical Guide.* Refer to www.oecd.org/dataoecd/42/34/42770261.pdf.

[12]National Science Board (2006). *Science and Engineering Indicators 2007.* Arlington VA: National Science Board.

[13]The author Co-chaired the second committee with Nigel Lloyd, representing Canada.

expected behavior is less ambiguous. In developing a case and compiling evidence, it is important to have standards that are easily understood by the investigators. And to prevent misconduct, the researcher should have a clear understanding of the expectations.

Another difference is how confidentiality is handled during an investigation. Some countries require full disclosure of individuals involved in the investigation. In others, like the United States, investigators may keep the investigative record closely held until it is adjudicated.[14] As a rule of thumb, one should assume that confidentiality operationalizes at the lowest standard of a participating nation. If one country's policies require early disclosure of allegations and accused, there is no practical way to hold information in a global inquiry or investigation. This does not mean that a country should violate its own requirements. It means that differences will occur and should be planned for in advance. In short, an investigation should be conducted with as much confidentiality as possible. The principle is to protect the innocent and to encourage the reporting of allegations.

The second GSF Committee agreed that all agreements governing international collaborations should be in writing and should address the policies and procedures that are to be used in the event there is an issue of research misconduct, the second GSF Report suggests boilerplate text that should be included in all collaboration agreements.[15] Such text needs to be agreed upon in advance of the project start. Once an allegation is made, the parties responsible for handling investigations should not waste valuable time negotiating the details of how to handle the investigation.

From a practical perspective, the investigative team should be experienced, and qualified. The procedures should have structure defining authorities and responsibilities. The scope and limitations of an investigation should be well defined, including definitions, standards of proof, and rules of engagement. The last addresses the management of the investigations. Specific issues include rules for confidentiality, independence of the investigators, handling of evidence, management of any conflicts of interest, and the sources and amounts of investigative resources.

---

[14]When an investigation is conducted by a federal entity the records are protected by the Freedom of Information Act and the Privacy Act.

[15]The boilerplate text lays out four points: standards of behaviors, reporting misconduct, cooperation in investigations, and acceptance of investigative conclusions.

An often-discussed issue is the potential problem of false allegations. Although a real threat, the experienced investigator can easily assess the seriousness of any allegation. The experienced investigator will initiate an inquiry stage to quickly assess whether there is sufficient concern to move forward into an investigative mode. Unless an entity, either governmental or private, has an established system to handle the investigation of allegations, the challenge of doing so can be overwhelming. It is these circumstances that invite spurious allegations. A well-developed and well-managed investigative process will not be distracted by false, malicious-intended allegations.

Two notions to be considered when developing structures to handle misconduct allegations are balance and detriment. The second GSR report discusses balance as a responsibility of investigators. Competing tensions within an investigation require judgments that may be unpopular. The example used is disclosure of individuals versus holding information confidential. It is important that the investigators consider the impact that balancing these tensions on the fact-finding purpose of the investigation.[16] No detriment deals with the presumption of innocence. No person should suffer until the facts are in. Also, no one alleging misconduct should suffer adverse consequences. Whenever possible a safe harbor should be available to persons bringing forth allegations. Sometimes called "whistleblowers," accusers must feel that the process will protect them. This challenge is also more difficult in a global environment because judgments cannot be controlled by the investigators. In the United States, when an investigation is conducted by a federal governmental entity, the investigation record may be kept confidential until the case is closed. The GSF report suggests that confidentiality should be maintained to protect those involved in an investigation. This protection includes the investigators themselves. They too need a safeharbor in which to operate. However, confidentiality should not be used as a gimmick to raise barriers during an investigation, or to jeopardize the health and safety of research participants.

The two GSF reports also recognize that the investigation, adjudication and appeal processes should have degrees of separation. This separation is important to recognize developing integrity structures because degrees of separation help assure fairness. The investigator should not be the adjudicator. In turn the adjudicator should not handle an appeal process.

[16]Boesz and Fischer, in Andersen and Steneck's *International Research Collaborations* (2010), suggest that striking a balance will not make all parties happy.

Investigations are simply fact-finding tasks. The accused should own an opportunity to check the facts established in the investigative record, and discrepancies should be recorded as part of the investigative report. Because the accused has been involved in reviewing the facts, an appeal step is not necessary at this juncture. However, adjudication and the decision on sanction or penalty are serious matters that need an opportunity for appeal.

Another international organization that has considered the research integrity challenges is the European Science Foundation (ESF).[17] In its 2010 report, Fostering Research Integrity in Europe, recommendations include adoption of a Code of Conduct.[18] The Report discusses core elements of a framework for research integrity governance and structures. These elements track the key concepts in the GSF reports. The ESF report emphasizes that each country must develop its own structure, agreeing with the notion that integrity structures must begin at home. At present, the US, Denmark, Norway, Finland, Australia, Canada and Germany are leaders among the small number of countries with established integrity structures.[19] This dearth indicates that there remains a critical need for continued work in developing integrity structures.

There is no internationally recognized entity that owns the responsibility for handling integrity issues. The advantage in this is that there is room for flexibility in developing structures that fit the needs of individual countries and the research institutions within these countries. The disadvantage is that the growth in international collaborations is outpacing the development of the integrity structures and this can lead to major difficulties for the researcher and the research enterprise when problems arise.

In summary, to assist entities in developing a credible investigative process, the two GSF reports, the ESF report, and now two World Conferences on Research Integrity have identified critical steps to be used in developing integrity structures. There is nothing mystical and nothing magical about developing effective, efficient research integrity structure. Good structures lead to good processes that in turn, lead to good outcomes. As you evaluate the various structures in place in individual nations, ask

---

[17]ESF led the promotion of research integrity in Europe in 2010 with its report. *Good Scientific Practice in Research and Scholarship.*

[18]European Science Foundation (2010). *Fostering Research Integrity in Europe.* Refer to www.esf.org/activities/mo-for a/research-integrit.html.

[19]ESF (2010). p. 10.

yourself the questions: Do they support research integrity? Can they handle allegations of wrongdoing? Are investigation and adjudication processes fair and timely? Are they transparent? Can the policies and procedures be harmonized to handle allegations of wrongdoing in international collaboration? The answers will help you assess how well the integrity systems support the research enterprise.

CHAPTER 2

STAKEHOLDER LEADERSHIP IN ADDRESSING RESEARCH INTEGRITY CHALLENGES*

Howard Alper

The general public, media, decision makers as well as the community of scholars, have become appreciably more aware of the need to address research integrity challenges. The control of research misconduct is essentially a regulatory problem, and we can learn much from the modern or "smart" regulation used by public institutions.

Self-regulation by stakeholders is an important instrument. Research integrity has many different components, and instrument choices should be tailored to the context. Overall, the key elements are instrument choice and implementation. Much can be gained from early and determined leadership and implementation by stakeholders.

It is crucial that leaders of universities, learned societies and associations, and other organisations, take an active role in establishing appropriate and effective research integrity policies through an outcome-oriented code — i.e. a combination of the best elements of a values-based perspective (encouragement) and a compliance-based perspective (rules and standards) to address issues of research misconduct. In terms of encouragement, it is essential to instill in our young, a value system that cherishes integrity. Starting to do so, at the university level, is far too late. Rather this needs to be achieved through education in elementary and high schools, and the young will also benefit by the development of a strong science culture. Furthermore, at this World Conference on Research Integrity, each of

*I am indebted to Professor Marc Saner, of the University of Ottawa, for his substantive contributions.

us here today, not only as stakeholder leaders, but as responsible family members, has a pivotal role to play at home in regards to inspiring children to adhere to an acceptable value system.

What is research misconduct? It is the violation of the standard codes of scholarly conduct and ethical behaviour in research. Examples include not only falsification and fabrication of data, and plagiarism, but also the misuse of research funds, and blatant conflict of interest in, for example, the adjudication of research proposals or papers submitted for publication. In other words, research misconduct applies to anything that violates "good research culture", and thus negatively impacts the credibility of the entire research community.

In establishing research integrity policies, we recognise that the scope of the endeavour is diverse (ethics contexts) and vast (numerous researchers on a global basis). There is an array of "regulatory" instruments that can be considered, and we should think in terms of a matrix, with the right tool for the right job — carefully selected, implemented, and evaluated. This can go a long way to assuring the responsible conduct of research.

In pursuing effective research integrity policies, we need to set the bar high.

"One of the trust tests of integrity is its blunt refusal to be compromised." (Chinua Achebe)

"Integrity is not a conditional word. It doesn't blow in the wind or change with the weather. It is your inner image of yourself, and if you look in there and see a man (person) who won't cheat, then you know he (the person) never will." (John McDonald)

Learned and professional societies are to be leaders on this issue, in addition to universities, industries, and governments. They should lead by example. Researchers who belong to such societies need to adhere to research integrity policies of societies concerned, their institutions where they are employed, as well as granting agencies or equivalent.

In addressing major violations of research misconduct, these policies often have a limited amount of formal oversight mechanisms. Only rarely is the conventional judicial process available. Criminal cases may result on an exceptional basis — e.g. where a possibility of harm may occur, such as in human health research.

Some learned societies and associations have rules which address important issues. For example, the Canadian Federation of Earth Sciences states that research integrity policies require clear guidelines on conflict of interest issues and quality control and quality assurance (QCQA). Concerning the latter, a detailed QCQA plan (with protocols for all aspects of the work) must be prepared, reviewed and agreed to before the research work begins, and has to be adhered to throughout the work project.

Stephen Pistorius, of the Canadian Association of Physicists, summed it up well.

"Learned and professional societies, as well as universities, have the responsibility to educate its members in research integrity and thus encourage its members to act ethically and with integrity, and be an example to is members through its own actions."

Setting the bar high for research integrity, and employing an appropriate mix of instruments, results in smart and determined stakeholder leadership. We can learn from practices in other jurisdictions. For instance, in sports, performance enhancing drug use is not a criminal act but it is regarded as professionally unacceptable. The profession monitors itself, and makes public the names of those who violate the code. Another example comes from the professional ethics of physicians. The Canadian College of Physicians monitors and takes actions against physicians who engage in unprofessional conduct (intimate relationships with patients).

In conclusion, effective stakeholder leadership described herein can prevent a level of misconduct that will trigger the creation of a crude external response (a "regulatory hammer"); maintain the trust and credibility by society in the research professions; and maximise the tailored, smart and effective use of a variety of instruments ranging from written rules to the fostering of a culture of integrity, resulting in fewer cases of research misconduct. The end result is likely better performance and a better public image.

CHAPTER 3

RESEARCH INTEGRITY CHALLENGES — A SINGAPORE PERSPECTIVE

Lee Eng Hin

1 INTRODUCTION

The ultimate goal of scientific research is the discovery of new knowledge that has to be verifiable so that the findings can be eventually accepted by the scientific community. In this competitive world, scientists are under pressure to "publish or perish". In recent years, there has been a proliferation of scientific journals and more recently e-journals and open access journals that encourage scientists to submit their work. In this new environment, safeguards must be put in place so that the published work has been sufficiently reviewed such that the information contained in the journals is of the highest standards of integrity.

2 BUILDING UP RESEARCH CAPABILITIES IN SINGAPORE

Singapore entered the arena of biomedical research rather late, but in the past decade it has systematically built up the infrastructure for basic science and translational and clinical research at the Biopolis[1] as well as in the universities, academic medical centres and healthcare institutions. There is now a critical mass of young as well as established scientists who have either been nurtured locally or in top overseas centres. In addition, Singapore has

[1]Biopolis is a biomedical research hub with co-location of A*STAR research institutes as well as private research labs in an area over 200,000 square metres.

strategically recruited many world-renowned senior scientists from abroad to help jumpstart our biomedical initiative. This systematic approach has earned Singapore the reputation of being a biomedical hub in Asia and has positioned Singapore to be a player in the world stage.

3 RESEARCH INTEGRITY — A HOLISTIC APPROACH IN SINGAPORE

Singapore prides itself to be a country with a very low tolerance for corrupt practices. In fact, it was ranked the 3rd least corrupt country in the world in 2009 by *Transparency International.* Singapore was ranked 2nd in the world for IP protection in the 2008 *IMD World Competitiveness Report.* Against this backdrop, Singapore has put in place many acts, regulations and guidelines to ensure best practices in patient care, biomedical research and clinical trials. Of particular importance are the Guidelines for Good Clinical Practice (GCP), Ethics Review Boards for care and research on animals and Institutional Review Boards (IRBs) for research involving human tissues and clinical trials. Singapore has also set up a Bioethics Advisory Committee to advise the Life Sciences Ministerial Committee on the ethical, legal and social aspects of life sciences research.

4 ENSURING RESEARCH INTEGRITY IN A*STAR

Recognising the need for the highest standards of ethics and research integrity, A*STAR, as a major funder of research, has put in place guidelines and processes to ensure that all researchers, young and old, are well acquainted with international best practices. The Biomedical Research Council of A*STAR has its own intramural research institutes and also funds biomedical research in the extramural community (universities, academic medical centres and healthcare institutions). Research integrity challenges occur in both the intramural and extramural community.

Intramural: Compromise in research integrity takes many forms, ranging from questionable research practices, such as failure to report conflict of interest to more serious infringements such as re-publication of published data in another journal, to outright research misconduct, such as falsification of data. As a preventive measure, A*STAR has developed a "Code of Best Research Practices" that is inserted in the Orientation Handbook for new recruits. This is also available on our Intranet. All staff are required to make a Conflict of Interest Declaration at the time of employment as well as on a yearly basis. Research Integrity Guidelines (including a Whistle-blowing Policy) has also been developed and readily available to the staff on the Intranet. These policies stress the importance of adhering to the Guidelines and the staff are encouraged to discuss issues openly with their peers and their supervisors.

Extramural: To ensure A*STAR grantees do not conduct irresponsible research, A*STAR provides grants only to institutions with reputable records and proper mechanisms to uphold research integrity and includes a conditional clause in the grant terms that allow A*STAR to terminate funding should the labs be found to have compromised research integrity. Another important area is the administration of the grant review process. As Singapore has a small population of 5 million people, it is not always possible to find grant reviewers without conflict of interest. We have thus instituted a two-tier grant review process with external (international) reviewers providing the input on international competiveness of the science and a local panel for final adjudication. Issues such as conflict of interest of the reviewers and duplication of grant applications are taken into account. A*STAR also holds international joint grant calls with a growing number of sister agencies from various countries. In these partnerships, prior agreement is made to ensure that the review processes are fair and transparent to both countries.

5 RESEARCH INTEGRITY CHALLENGES IN TRANSNATIONAL COLLABORATIONS

The most innovative research flows out of transnational and cross-border research collaborations especially when the expertise is complementary. To this end, Singapore has established collaborations with many countries in North America, Europe, Asia and Australasia. In these transnational collaborations, certain ethical and research integrity issues may arise. Harmonisation of ethical standards is one of the ways to ensure the viability of such collaborations. An example of this is the development of Guidelines for Human Embryonic Stem Cell Research by the International Society for Stem Cell Research (ISSCR) which was achieved through an international task force. The cross-border transfer of research materials or clinical samples needs to be handled diligently and with proper agreements and IRB approvals. Prior agreements on authorships, ownership of data, etc. has to be followed. Following the ICMJE authorship guidelines would be an acceptable practice.

6 RESEARCH INTEGRITY CHALLENGES IN INDUSTRY COLLABORATIONS

As more industry collaborations are entered into, scientists must be aware of certain industry requirements such as withholding publications until patents are submitted and certain publication practices that are designed to enhance the quality of the data or to improve the marketing of the product. One way to ensure more parity in decision-making would be to enter into co-funded collaborative projects where the researchers co-share the risks and thus the benefits from the research outcomes. This may allow more control of the data and subsequent publications. It is important to enter into such projects with prior agreements. In reporting such research work, it is important to make conflict of interest declarations upfront.

7 BUILDING AN ETHOS

As the research community grows and more and more young scientists join the ranks, it is imperative that a sense of ethics and research integrity is inculcated into these young and impressionable minds. To this end, A*STAR has established a set of clear guidelines that are available in hard copy as well as on the web for all the scientists. In addition, regular workshops using relevant case studies are organised. These workshops provide real-life examples for discussion on research integrity. Attendance is mandatory for all ranks in the A*STAR family. Open discussion is encouraged and feedback to fine-tune the guidelines are appreciated. In this way A*STAR hopes to promote best practices in research and reduce the likelihood of breaches in research integrity practices in the intramural and extramural research community.

CHAPTER 4

EUROPEAN SCIENCE FOUNDATION AND RESEARCH INTEGRITY

Ian Halliday

The European Science Foundation (ESF) is a member organisation consisting of the main funding agencies and research performing organisations of Europe together with many of learned societies.

In this chapter, I would like to address two main topics. First, I describe the activities in research integrity that ESF Member Organisations have shared and discussed together. Second, I would like to express the idea that we are in real danger of allowing the integrity debate to be too strongly focused on our problem cases and challenges and thus allow the innate strength of the scientific endeavour to be undermined. Given the increasingly high profile of science, I also suggest that the scientific community needs to play an active role in articulating and organising our relations with the non-scientific world outside science.

In the area of Research Integrity (RI), ESF has enabled discussions across a wide spectrum of European organisations through its Member Organisation (MO) Forum. More details can be found in the two ESF publications referenced at the end of this section.

1 BACKGROUND TO THE ESF MO FORUM

ESF is very proud to have played a founding and leading role in the Lisbon Conference. Since then, and based on an earlier ESF study, its activities have developed from this engagement.

Following the first World Conference, ESF held a workshop in Madrid in 2008 after which the MO Forum was established. It consists of four

Working Groups addressing the following topics: raising awareness; code of conduct; setting up national structures; and furthering research on RI. The Forum brought together 34 of its 78 Member Organisations plus the participation of other European groups including the European Federation of National Academies of Sciences and Humanities (ALLEA), the European Universities Association (EUA) and the European Network of Research Integrity Offices (ENRIO). It held several workshops, one of which was devoted to the topic of education and training in research integrity organised jointly with United States Office of Research Integrity (US ORI) and the National Science Foundation (NSF).

2 SCOPE EUROPEAN CODE FOR RESEARCH INTEGRITY

One of the aims of the Forum was to discuss and formulate a European Code for Research Integrity (CoC). This is not a body of law, but rather a canon for self regulation. It applies to research in all sciences and fields of scholarship: natural and life sciences, social sciences and humanities. The CoC confines itself to standards of integrity while conducting research, and does not consider the wider socio-ethical responsibility of scientists and scholars.

Outlined below are the main contents of the Code:

Principles of scientific integrity

Researchers, research institutes, universities, academies and funding organisations should commit themselves to observe and promote principles of scientific integrity. These include:

- Honesty
- Reliability
- Objectivity
- Impartiality and independence
- Open communication
- Duty of care
- Fairness
- Responsibility for future science generations
- Research employers have a responsibility to ensure that a culture of research integrity prevails.

The Code sees the two most serious violations as being Fabrication and Falsification with a third category of misdemeanours being:

- Infringement of intellectual property rights (including plagiarism)
- Not (or Improper) dealing with violations of integrity
- Petty misdemeanours ("adjustment" of a figure, cutting a corner, trimming of data, omitting an unwelcome observation): unacceptable and whilst not giving cause to a formal charge they should be reprimanded and corrected.

European Examples of Implementation

Along with the general ideas of Integrity, there is also the question of how the different models are implemented. Which organisation takes final responsibility? The table below gives a rough indication of the spread of responsibilities.

Country	Agency/ Professional Body	Local with National oversight	National
Croatia			√
Denmark			√
Finland		√	
France	√		
Germany	√		
Hungary		√	
Ireland	√		
Latvia	√		
Netherlands		√	
Norway			√
Poland	√		√
Spain	√		
Sweden		√	
Switzerland	√		
UK		√	

Research Integrity Governance in practice

The MO Forum has examined and tabulated the various governance structures that exist in Europe. Firstly, there are examples where funding agencies or national professional organisations take responsibility.

Agencies/Professional bodies

Country	Organisation	Role	Scope/ Limitations
Ireland	Health Research Board	Advisory	HRB funded research. Rely on institutions to have mechanisms in place for dealing with misconduct
Germany	Deutsche Forschungs-gemeinschaft (DFG) Ombudsman's Service + Max Planck Society	Advisory and investigatory	Require institutions to have mechanisms in place for dealing with misconduct. Investigatory in DFG or MPS funding only
Switzerland	Swiss Academies of Arts and Science in collaboration with Swiss National Science Foundation	Advisory, oversight and investigation	All publically funded research. Local investigation supported by legislation

Secondly, there are examples where the responsibility is local to a university or Research Institute with national oversight or assistance.

Local with national oversight

Country	Organisation	Role	Scope/ Limitations
Netherlands	National Board on Scientific Integrity (LOWI) (Secretariat in National Academy of Arts and Sciences)	Advisory, oversight and appeals	All universities and institutions under LOWI umbrella. Voluntary. Private sector not included
UK	UK Panel for Research Integrity in Health and Biomedical Sciences (UKRIO)	Advisory and oversight	Covers only health and biomedicine — public and private sector. Proposed National Advisory Body for all disciplines
Finland	Finish National Research Ethics Board	Advisory and appeals	University. Voluntary sign-up

Finally, there are examples where a national body, separate from the funding agency, has responsibility.

National Body

Country	Organisation	Role/ Mandate	Scope/ Limitations
Norway	National Commission for the Investigation of Scientific Misconduct	Advisory and investigatory. National legal jurisdiction	All public and private sector institutions
Denmark	Danish Committees on Scientific Dishonesty	Investigatory, may proactively take cases. National legal jurisdiction	All public sector institutions and universities
USA	Office of Research Integrity	Advisory, oversight, investigatory, regulatory. National legal jurisdiction	Public Health Service institutions worldwide. Does not cover private sector

3 SCIENTIFIC INTEGRITY IN THE WIDER CONTEXT

In this section, I want to express my own personal opinions rather than represent the activities and position of the ESF itself.

In a previous life, I was, for seven years, the CEO of one of the UK Funding Agencies. This meant that I operated on the boundary between the science community and the government. The area of science was "Big Science" as exemplified by particle physics and astronomy. I came away with one over-riding message about this relationship. It is built on TRUST of a visceral nature.

This TRUST operates at a whole series of levels. The funding agency must prove financial capability and credibility. The scientific community must show discipline both in what it promises and in describing the limits of its knowledge. The temptation to promise instantaneous medical cures and/or immediate industrial breakthroughs is sometimes irresistible. This is just as big a threat to the credibility of the science community as any fraud.

The route by which such claims are propagated often involves the press and media. There is, therefore, a challenge to properly engage with the individual reporters who may not have scientific backgrounds. Many scientists confuse outreach and education with salesmanship for their particular scientific specialist field.

Moreover, recent examples demonstrate the ambiguity and threats not from fraud but from accusations of lack of integrity. An obvious example is the "Climategate" case involving the University of East Anglia in the UK. This involved the small Climate Research Unit, engaged in tree ring dating research and its implications for global warming. The discussions at the very high-level political Copenhagen conference on climate change were affected by the claims and counter-claims surrounding this work. The discussion was driven by stolen e-mails and fanned by tardy Freedom of Information openness. Following the stories in the press, several independent reviews were instigated led by Lord Oxburgh looking at research integrity of the science itself and by Sir Alastair Muir-Russell investigating the process. To an interested outsider these reviews are clearly exercises in "shutting stable doors after the horse has left." They are not seen as totally credible, even though they are independent. However the main message of the reviews is that the relationship between science and the wider political environment was not well handled either by the university or the general scientific community.

But the science community is clearly left with a challenge. How did this happen and what needs to be done to correct or properly defend the scientific system?

4 SCIENCE AND TECHNOLOGY GROW

It is clear that, over the past two centuries, the impact of science and technology has grown beyond all predictions. Two recent examples make this clear. One is the growth and impact of the digital IT revolution, particularly since the Web was invented to seamlessly join computers, which has impacted on politics, security, industrial wealth employment, etc. Similarly, the present and looming future effects of climate change will have major impact on politics and governance. Both easily span national borders.

It is not clear that science has adjusted to these changes. Thus, science is no longer an introspective subject with free, lone scientists as the archetypal norm.

Most areas of science have policy and societal impact. Sometimes, this is in the short term, sometimes it is over 50 year timescales. But most areas have not accepted the consequences of such changes. This new world clearly involves engagement and clear understanding with other actors.

So science needs to be very clear about the interactions with industry, media and politics. It is my view that this interaction must become much more pro-active and institutionalised than at present. The on-going revisions of the methodologies and politics of the UN Panel on Climate Change seem to me to show a path many others will be required to follow.

5 ... SCIENCE AND TECHNOLOGY GROW BUT

If we follow the logic of the above then, as part of the integrity and TRUST debate, the methods of traditional science may have to change. Thus, the isolated researcher of the classical science model is very vulnerable if exposed to the media in a crisis situation. The supporting environment will become increasingly important. Universities, for example, will have to become much more media "savvy". It is very noticeable that big science organisations such as CERN or NASA can engage and control media interactions in a way difficult to achieve for smaller organisations.

A challenge for the science community is to engage with other communities in denial about scientific certainties. Do we engage at a political level or remain above the fray except for brave individuals?

At the level of the individual scientist, a transparency and record-keeping environment analogous to that required by the patent situation will become much nearer the norm.

The growing move to publication with accompanying mandatory data archives will also have major consequences that the science community is not totally aware of. Big science, in particular, astronomy is already aware of the value of well-curated data sets but is also very aware of just how large an effort is required to achieve this.

The underlying question is how the science community should engage with politicians and the public over future consequences of scientific advance. Sometimes this engagement is driven by science asking solely for extra government money. Or, there are great discussions following some disaster.

The question remains open as to organising information flow in a form useful to both sides. This flow should have the true imprint of integrity and trust on both sides. Many organisations are attempting to open such channels of trustworthy communication. Some examples are Cambridge University and the Royal Society of Edinburgh opening channels to the UK and Scottish Civil Service in a non-crisis situation. This strikes me as an area where visibility of different national efforts would be useful although clearly the systems will match different national systems.

6 ACKNOWLEDGMENT

I would like to thank Vanessa Campo-Ruiz of ESF for preparing much of the material in the first section.

References

1. Details about the ESF MO Forum on Research Integrity together with reports and papers may be accessed from http://www.esf.org/activities/mo-fora/research-integrity.html.
2. *Fostering Research Integrity in Europe — Executive Report A* report by the ESF Member Organisation Forum on Research Integrity (June 2010), ISBN: 978-2-918428-15-2.
3. *Good scientific practice in research and scholarship.* ESF Science Policy Briefing No. 10, (December 2000).
4. *Research Integrity: Global Responsibility to Foster Common Standards* (Mayer and Steneck (eds.)) ORI–ESF Science Policy Briefing No. 30 (December 2007).
5. *Stewards of Integrity — Institutional Approaches to Promote and Safeguard Good Research Practice in Europe: An ESF Survey Report* (May 2008), ISBN: 2-912049-82-2.

CHAPTER 5

FRANCE: HOW TO IMPROVE A DECENTRALIZED, AMBIGUOUS NATIONAL SYSTEM

Jean-Pierre Alix

1 CURRENT SITUATION IN THE 2000s

Following my return to France after the OECD–GSF Workshop in Tokyo (2007), I noted that nothing was being done to promote research integrity and to deal with cases, except in one University (Lyon) and one Research Organisation (INSERM). In detail, the situation was as follows:

- There was no standard methodology for handling cases ("under the carpet" behaviour) resulting in unfair and unequal treatment of cases;
- There was no serious knowledge about Fraud and Research Integrity (RI);
- There were no serious comparisons with other countries;
- These factors implied future problems with public trust.

The challenge was how to address this in order to reach a better situation. In other words, we needed to avoid what had been experienced in many other countries and to avoid any increase in fraud and misdemeanours. This necessitated establishing a realistic system and procedure with rules or guidelines.

But there were some barriers and limitations.

- Many French scientists think that fraud is a very minor phenomenon, so it is not important with some others also considering it is a taboo topic;
- Many considered that research integrity should be dealt with within research institutions themselves before any publicity results;

- Finally, there was strong advice "not to transform a value based action into a bureaucratic system (creating its own 'industry')". Therefore, one had to be critically aware of the argument about being seriously aware of "value for cost and time".

2 MINISTRY OF HIGHER EDUCATION AND RESEARCH LOOKS FOR A BETTER SITUATION

The goal was explicitly defined to deal with RI, base the processes on prevention and education, with publicly fair procedures (guidelines) to deal with cases. My mandate was to investigate and make proposals. The approach had to be seen as non-bureaucratic, encouraging the scientific community to consider fraud and misconduct as a real problem. Four steps were defined with a duration of at least two years assigned for the process:

1. An international review to define what is misconduct and fraud;
2. A national survey to create some statistics;
3. A report to ministry including recommendations;
4. Implementation.

3 CURRENT RESULTS AS AT MID-2010

The survey was conducted in 2009 with the main findings as follows:

- There is less than one case per a year by institution on average although others, notably INSERM, declare up to 6 cases per year;
- There is already a legal base for action;
- Prevention and education is weak;
- There is an acceptance of the need for clear rules and procedures in the future coupled with an acceptance for a national appeal system;
- In general, the survey indicated a good will within the research community to improve the RI system.

A Report has now been made to the Ministry which covers topics such as a description on how fraud arises including the pressures on individuals to publish, to innovate, and to conduct outreach and on laboratories to deliver results which may create situations where rules are being transgressed. It proposes a Charter based on science values with a definition of main misconducts frauds (Falsification, Fabrication of data, Plagiarism). It includes a comparison of the French national situation with that prevailing in other countries in terms of law, processes, number of cases, etc. It then provides recommendations for prevention of misconduct and the promotion of good practices mainly by Universities, through Guidelines. Other recommendations deal with institutional procedures (whistleblowers, location of allegation, protection of both parties by confidentiality and fairness of procedure, legal base for expertise and decision, conclusions, decision by employer). The creation of a national Appeal system is a key recommendation. The Report also envisages having a national annual report on research integrity. Finally, it proposes a plan for future implementation. This Report is currently under discussion by the Academy of Sciences, Ethics Committees, leading research personalities, etc., in order to both disseminate the questions to be solved and to develop support and allies in the research community.

4 NEXT STEPS 2010–2011

The Report will now be disseminated to institutions to increase awareness after the survey, to be followed, in autumn 2010, with a Workshop dedicated to governance of RI by the Ministry, Funding Agencies, Research Organisations and Universities. It is hoped that Rules governing RI will be adopted in the first semester of 2011 by each institution. Finally, it is hoped to organise a national public conference in 2011.

5 RECOMMENDATIONS FOR A SUCCESSFUL APPROACH

There is a need to focus on a very few strong arguments. For example, RI is an integral part of research activities and has to be promoted as such the law is not always necessary. Fraud and misconduct in science is a failure to oneself, to colleagues and to the public. Harmonisation is necessary for two reasons: one is the intricate and complex funding situation in laboratories with many reporting and accounting responsibilities, with these being coupled with frequent international collaborations. Any system created has to respect national and cultural traditions so that action should possibly be limited to the main frauds.

6 ABOUT THE PROCESS TO START A SYSTEM

The "taboo" situation of fraud is the main obstacle: "Why talk about such behaviours?". An open discussion is needed, because fraud is considered as a major and inexcusable misconduct by most scientists, even though it is an infrequent occurrence. A body which acts to create a better situation, of necessity, has to find allies within the scientific community. A step by step approach is needed because many actors are involved. The most efficient approach is to search for a common basis. This takes time for information dissemination, for proposal formulation and discussion and then for implementation.

7 SUMMARY

As a cultural behaviour, RI needs attention from national authorities (national and international coherence is needed), but the main responsibility lies in laboratories and especially among scientists themselves as individuals.

In a "no man's land" situation or in an ambiguous atmosphere, the goal has to be to establish clear common rules, and, even more difficult, to have them adopted.

"Give time to convince and avoid forced decisions" is a strong guideline, because a bad system, even if adopted, does not really work. Even an efficient system has to be discussed and promoted with every coming generation, as well as within laboratories and by scientists themselves.

A dedicated law might be promulgated, but is not evident that it will work, because basic laws for honesty, intellectual property, etc. already exist in many countries and are frequently breached.

CHAPTER 6

RESEARCH INTEGRITY IN THE CANADIAN CONTEXT

Ronald Heslegrave

1 OVERVIEW

Research plays an important role in all societies. Decisions, policies, regulations, inventions, and innovations are generally based on some form of research. Research is essential for the advancement of social, medical, public policy and economic agendas. Given this dependence on research, the *integrity* of research is vital. Failure of the system to support the integrity of the research enterprise threatens progress at all levels of societal development. Decision makers, the public, and even researchers themselves, must have confidence in research, how it is conducted, and the subsequent findings.

2 THE CANADIAN PERSPECTIVE

There are a number of organizations or initiatives in Canada that have been instrumental in attempting to forward the research integrity objective. The Natural Sciences and Engineering Research Council (NSERC), Social Sciences and Humanities Research Council (SSHRC), and the Canadian Institutes of Health Research (CIHR) are the three federal agencies responsible for the public funding of university-based research and training in Canada. In Canada, these three funding agencies, commonly referred to as the "Tri-Council", have been very active in setting the highest standards of integrity in research and scholarship. In 1994, the first draft of the Tri-Council Policy Statement on Integrity in Research and Scholarship (TCPS–IRS) was created. This policy statement is essentially the centerpiece of the current research integrity system. Given that many institutions rely on at least one of these organizations for research funding, many individual institutional policies share some of the key methods and approaches outlined in the TCPS–IRS.

In 2008, following a number of media reports regarding research misconduct and the misuse of funds by NSERC-funded researchers, the Minister of Industry requested an internal review from the Tri-Council and the Association of Universities and Colleges of Canada (AUCC). While the review concluded that the overall approach to research integrity was essentially sound, not long after in 2009, a relatively new organization in Canada, the Canadian Research Integrity Forum,[1] commissioned a report to examine the state of research integrity policies in Canada. The report, published by consulting firm, Hickling, Arthurs and Low (HAL), concluded that the research integrity system in Canada would benefit from strategic strengthening.

3 KEY GAPS

The HAL report highlighted the fact that Canada does not have a harmonised system of research integrity. Rather, Canada's system can be characterised as a "hub and spokes" model, with the Tri-Council maintaining a central position, coordinating and mandating the development of the institution-specific research integrity and misconduct policies. As institutions are charged with the development and enforcement of research integrity policies under general guidelines from the Tri-Council, the interpretation of such polices differs from institution to institution and also across disciplines. Thus, Canada's system remains segmented with no national definition of misconduct or research integrity.

It should be noted that the Tri-Council is only *one* funding option and *one* facet of the research integrity debate in Canada. More and more research is being funded privately, and in some cases a mixture of public and private funds is required. This adds another layer of complexity as granting agencies often partner with industry to enhance innovation and promote the translation of research into practical societal benefits. However, research

[1]Formerly the Canadian Research Integrity Council (CRIC), the CRIF was created in 2007 to bring together government and non-government organizations (NGOs) that share a common interest in strengthening research integrity in Canada.

integrity policies and sanctions only apply to the use of public funds used in research while private funds are regulated through contractual obligations.

With a global perspective on research integrity and its emerging importance, coupled with the key gaps in the oversight of research integrity as identified in the 2009 HAL report, it is apparent that Canada, should strengthen its current system in a manner that is consistent with and integrated into the global perspective. Urgent attention to these issues is seen as a scientific and ethical imperative.

4 THE EXPERT PANEL'S CHARGE

In 2009, the Council of Canadian Academies was asked by the federal government to conduct an assessment to examine the key research integrity principles, procedural mechanisms, and practices in Canada. The Council of Canadian Academies is a not-for-profit organization that supports independent, science-based, expert assessments that inform public policy development in Canada. Council assessments are conducted by multidisciplinary panels of experts from across Canada and abroad. The Expert Panel on Research Integrity was formed in order to address the following question and series of sub-questions:

What are the key research integrity principles, procedural mechanisms, and practices, appropriate in the Canadian context, that could be applied across research disciplines at institutions receiving funds from the federal granting councils?[2]

What definitions do research institutions (i.e. post-secondary institutions receiving funding from the granting councils) employ for research integrity in Canada and how could these approaches be made more uniform?

How would the Canadian definition differ from that of other countries, including the United States and why? How do we align the approach used for research integrity, by the granting councils and the Canadian post-secondary institutions, with that of leading countries and emerging global standards?

---

[2]Granting Council's refers to the three organizations comprising the Tri-Council: the Natural Sciences and Engineering Research Council (NSERC), the Social Sciences and Humanities Research Council (SSHRC), the Canadian Institutes of Health Research (CIHR).

What actions would be considered to constitute research misconduct in a Canadian context?

In light of a clear definition of research integrity, what are the roles and responsibilities of those involved in research (including researchers, scientists and research and academic institutions funded by Canada's granting councils) to uphold this definition and the key principles and practices, including roles and responsibilities for education?

How could a common research integrity definition foster a research culture of high ethical standards and instil public confidence?

5 THE ASSESSMENT PROCESS AND UPCOMING REPORT

The Expert Panel's report will examine the integrity of research in Canada, offer a definition of "research integrity", and provide some insight into acceptable research practice and procedures. During its deliberative phase, the Panel heard from many stakeholder groups including the Associations of Universities and Colleges of Canada, the Canadian Association of University Teachers and the Association of Canadian Academic Healthcare Organizations (ACAHO), who were invited to give presentations about their organizations and their views regarding research integrity.

In June 2010, the Panel's report was completed and underwent stringent and anonymous peer review by some 11 reviewers. Following the assessment of the reviews and incorporating the concerns of the reviewers into the final report, the Panel officially signed off on the report in early July 2010.[3] It is anticipated that the report will be made publicly available in the fall of 2010.

All Council reports are available in English and French in print and electronic format. To download assessment reports or receive notification of report releases, visit the Council's website www.scienceadvice.ca.

[3]At the time of the World Conference of Research Integrity in July 2010, the Panel's report was undergoing formal publication, including translation into French. The report was officially released in October 2010.

CHAPTER 7

RESEARCH INTEGRITY IN NEW ZEALAND

Sylvia Rumball and John O'Neill

New Zealand has no comprehensive approach to research integrity. Instead there is a complex web of legislation, codes, ethics committees (some of which are statutory) and Crown entities.[1] General legislation impacts on the conduct of research,[2] as well as the Treaty of Waitangi 1840 which concerns the relationship of the Crown with the indigenous people, the Māori. Separate legislation[3] requires all tertiary education institutions to demonstrate and maintain the highest ethical standards, and to permit public scrutiny of these.

The majority of research in New Zealand falls outside this framework. Universities have responded by developing institutional codes and ethics committees. Informal national discussions as to whether a comprehensive

[1]Animal Welfare Act 1999, National Animal Ethics Advisory Committee and Institutional Animal Ethics Committees; Hazardous Substances and New Organisms Act 1996, Environmental Risk Management Authority and Institutional Biological Safety Committees; Human Assisted Reproductive Technology Act 2004, Advisory Committee on Assisted Reproductive Technology, Ethics Committee on Assisted Reproductive Technology; Health Research Council Act 1990, Code of Health and Disability Services Consumer Rights 1996, Operational Standard for Ethics Committees, Health and Disability Ethics Committees, Institutional Human Ethics Committees, National Ethics Advisory Committee and Health Research Council Ethics Committee.

[2]Such as the Human Tissue Act 2009, the Privacy Act 1993 and the Protected Disclosures Act 2000.

[3]Education Act 1989.

national code could be developed, as in Australia[4] and Canada,[5] failed. To date, attention has focussed on establishing procedures for ensuring responsible conduct of research in particular areas rather than development of a comprehensive research integrity framework. This situation is similar to that reported for UK higher education institutions.[6]

The emerging literature on scientific integrity identifies three levels for attention: personal, collective and institutional. To date attention in New Zealand has centred predominantly on the personal level. However, the prevailing culture, both at institutional and collective level, is an important determinant of the behaviour of individual researchers, particularly those new to research. And here, there is potential for conflict.[7]

The degree to which the more general concepts of research integrity are embedded at all levels (governance, leadership, senior management and research staff), is unclear. For example, there is rarely an explicit policy to inform decisions on research revenue and investment.[8] Few breaches of the framework reach the public domain, but no data is available concerning the numbers and types of institutional breaches of research integrity. There is

---

[4]Australian Government, National Health and Medical Research Council (2007). Australian Research Council, *Australian Code for the Responsible Conduct of Research.*

[5]Canadian Institutes of Health Research, Natural Sciences and Engineering Research Council of Canada, Social Sciences and Humanities Research Council of Canada (1998). *Tri-Council Policy Statement: Ethical Conduct for Research Involving Humans* (with 2000, 2002 and 2005 amendments).

[6]Council for Industry and Higher Education (October 2005). *Ethics Matters: Managing Ethical issues in Higher Education.* The Council.

[7]Sulston, J. (December 2008). *Science and Ethics.* The Biochemical Society.

[8]Commission of the European Communities. *Commission recommendation on the management of intellectual property in knowledge transfer activities and Code of Practice for universities and other public research organisations.* (C(2008) 1329), The Commission, April 2008. Retrieved 7 October 2009 from http://ec. europa.eu/invest-in-research/pdf/ip_recommendation_en.pdf. Gazette Oxford University. *University policy on socially responsible investment*, The University, March 2008. Retrieved 7 October 2009 from http://www.ox.ac.uk/ gazette/2007-8/weekly/190308/notc.htm#7. Ref University of Bath. *Institutional code of ethics,* The University, undated. Retrieved 7 October 2009 from http://www.bath.ac.uk/ vc/policy/ethics.htm. University of Manchester. *The University of Manchester Policy for Socially Responsible Investment*, the University, undated. Retrieved 7 October 2009 from http://www.campus.manchester. ac.uk/medialibrary/policies/ socially-responsible-investment-policy.pdf.

no requirement for public reporting of "sentinel events" as occurs in the public health sector.

Generally, where staff are involved, universities decline to release any information on the grounds that it is a "personal employment matter". Furthermore, we are not aware of any institutions which evaluate the integrity of their research environment using a process of self-assessment and external peer review.[9]

1 RECOMMENDATIONS

Attention should not be focused solely on personal research integrity but also directed to the group and institutional level in an integrated manner. Good behaviour needs to be modelled in both governance and management.

Leaders need to champion the development of an institution-wide culture which explicitly embraces integrity in all activities and, specifically in research, a commitment to self-review in order to maintain an informed balance between open and commercial science imperatives.

There is also a need for greater transparency concerning the incidence of breaches of research integrity via national statistics, publicly reported on an annual basis.

In an era when collegial norms and disciplinary traditions are under increasing threat, institutions need to devote continually resources to the education of staff, and to maintain norms of meaningful discussion and consistency of decision-making.

[9]Institute of Medicine National Research Council (2002). *Integrity in Scientific Research.* Washington, DC: National Academies Press.

CHAPTER 8

CHALLENGES ENCOUNTERED BY THE SWISS ACADEMIES OF ARTS AND SCIENCES WHEN INTRODUCING CONCEPTS FOR PROMOTING SCIENTIFIC INTEGRITY

Emilio Bossi

I would like to thank the organisers for the opportunity to give this talk. Before I start with the subject, let me briefly tell you what the Swiss Academies of Arts and Sciences are: they are an umbrella organisation of the four scientific academies and represent their research-promoting institution, which edits guidelines and directives for research conduct and ethics and which is felt as moral authority in the domain.

I was asked by the organisers to discuss challenges we were and are confronted with in Switzerland when trying to promote and to maintain high standards for integrity in research and to present our responses to those challenges. I don't have the task to speak about definitions, the main topic of this session.

So, I have chosen two of the challenges we experience in Switzerland: First, the underestimation of scientific misconduct by the scientific community, and, second, the scepticism and mistrust towards academic self-control by the community.

The first challenge is the underestimation of scientific misconduct by the scientific community. Scientists believe scientific misconduct to be very rare and that it would never occur in their own institution. This attitude is absolutely understandable, since for most scientists a responsible scientific conduct is self-evident.

Based on this attitude, however, many scientists believe that a special promotion of scientific integrity is not necessary, and that people promoting scientific integrity in an organised and formal way, for instance the Academies, are at best utopian theorists, or, even worse, missionary intruders who only increase the already heavy administrative burden of scientists.

The second challenge is the scepticism and mistrust towards academic self-control by the *community*. Parts of society, several politicians, a fair number of civil servants and almost all of the media would not accept that the scientific community handles suspicions of scientific misconduct itself, since self-interest, favouritism, and corruption will not allow the scientists to objectively handle such cases. This attitude is also understandable, because there are many examples of improper behaviour, of course not only in science.

1 SCEPTICISM AND MISTRUST TOWARDS ACADEMIC SELF-CONTROL BY THE COMMUNITY

How do we try to solve the first challenge? In order to correct the assumption that misconduct is not an entity, we see three ways:

1. Raising awareness about misconduct.
2. Training in responsible scientific conduct.
3. Installing national statistics about the occurrence of misconduct.

For the purpose of raising awareness, the Academies have elaborated a brochure on principles and procedures concerning scientific integrity, which is also available online (www.swiss-academies.ch). This text has been distributed to all research institutions in Switzerland, mainly to the universities. Among other contents the text strongly suggests the introduction of a formal integrity organisation in each research institution. This organisation is now established with minor modifications at most Swiss universities. Besides avoiding unprepared and inappropriate reactions to misconduct allegations which discredit the university and science as a whole, the mere presence of such an organisation increases the awareness of scientific integrity at the research institution.

Training students and young scientists in Responsible Conduct of Research is one of the ways to increase awareness in Research Integrity. For that purpose, the Academies have elaborated a basic lecture, an introduction to the subject which we have put on the internet and that can be used by the research institutions. We also offer to give this basic lecture ourselves. This opportunity has been used several times. Whether these activities help to reduce the occurrence of misconduct, we do not know. We assume, however, that a short formal introduction to the subject of integrity and misconduct should be given, but that it is even more important that the subject is repeatedly treated in a more informal way in discussions within the research groups. Our experience with teaching is still small and I am happy to learn more about the subject at this conference.

As for instituting national statistics, we are far from being successful. Research institutions are reluctant to declare their cases of misconduct, even anonymously. They do investigate suspicions, they often come to a conclusion, they also draw consequences — but they are not eager to inform even the scientific community, because they fear a loss of reputation for their institution. At the time being, I do not have a clear suggestion for solving this problem. What we can do is to divulge the statistical findings of others as they appear in the literature. But then, these data might not be applicable to our own national situation.

The next step we plan to take is to bring together the persons responsible for scientific integrity of the Swiss universities to discuss the problem. However, as these specific persons are positively biased towards research integrity, they usually do agree with the necessity of a national statistic. The problem lies on a higher hierarchic level at the universities and we have to figure out how to convince these persons. Here again, I hope to learn from other nations' experiences at this conference.

2 SCEPTICISM OR EVEN MISTRUST OF THE COMMUNITY TOWARDS ACADEMIC SELF-CONTROL

Due to this mistrust, politicians and civil servants might want to take the investigations away from the scientific community and to appoint governmental panels for handling cases of misconduct.

The Swiss Academies of Arts and Sciences strongly believe in the own responsibility of the academic community in matters of scientific integrity. Self-control and self-regulation are inseparably connected with this responsibility. Thus, management of suspected misconduct is a task for the research institutions themselves. However, we have to consider that the academic research institutions are part of society and are supported by governments. We have to accept the mistrust as a given fact and be open for collaboration with governmental institutions. External experts can and should be invited to collaborate in misconduct investigations, but the responsibility has to remain within the scientific community. It is our belief that restoring or strengthening mutual trust, as well as clear mandates are basic preconditions for scientific self-regulation.

In order to solve this second challenge, that is, to gain the trust and the collaboration of governmental institutions and at the same time to create an efficient mechanism to promote scientific integrity, we have chosen the following procedure. Maybe you can use some hints for your own purposes.

Most importantly, the Academies have taken the initiative in their own hands and did not wait for an invitation or ask for a mandate. A working group of the four scientific Academies has worked out the above mentioned principles and procedures for promoting scientific integrity. This self-initiative made it possible to work efficiently and with professionalism. Once we had formulated our text, we organised a hearing in which the interested parties, namely the universities, founding agencies, politicians, governmental institutions, research representatives of the industry, and the media were involved. We then finalised the text paying attention to the input the participants had given us. The text was then made public. Then, but only then, did we ask for a meeting with the Secretary of State responsible for Education and Research in which we presented our efforts. As a consequence, the Academies were endowed with the responsibility of networking the Swiss institutions that deal with scientific integrity and to represent Switzerland internationally in this field.

Furthermore, our role as one of the promoters of ethics in research is now stated in the national Swiss law on research (Art. 9b), as is the license to inflict administrative sanctions against offences of good research practice in our domain (Art.11a). Other offences remain a matter of punitive law and civil courts.

A gain in trust has been achieved, but we are aware that trust is not completely restored everywhere in the community. This is one of the main reasons why we have to permanently pursue our efforts to promote scientific integrity.

CHAPTER 9

THE INTEGRITY OF RESEARCHERS IN JAPAN: WILL ENFORCEMENT REPLACE RESPONSIBILITY?

Tohru Masui

Scientific integrity and research ethics have been essential issues of scientific research. As a research bioresource manager and a researcher I have been observing the way in which these issues have played out in the course of establishing guidelines on research conduct for biomedical sciences in Japan. In this paper I would like to discuss the behaviour of researchers in the course of the establishment of research ethics guidelines on human materials and information, and during the passing of data protection legislation in Japan. These cases show the role of external regulatory frames on the behaviour of the research community and researchers. They also demonstrate that the nature of difficulties in ensuring integrity among researchers depends on the form and nature of external regulation.

1 DISCUSSION ON THE RESEARCH ETHICS GUIDELINES

In Japan, substantial discussion on the research ethics guidelines for biomedical science especially on the use of human materials and information started in 1998, and was initiated by the Ministry of Health and Welfare. The discussion started abruptly and the research communities were not prepared. Members of the Kurokawa Committee, chaired by Dr. Kiyoshi, Kurokawa who were discussing medical research practices claimed that informed consent was already obtained as part of standard research practices. However, a survey conducted a few years later (Yasuhara and Kurata, 2002) on the same medical research areas reported that less than half of the laboratories were seeking informed consent for the use of leftover clinical samples. This shows us how the top rank research leaders were unaware of the practices of their colleagues. It also raises suspicions about the validity of the discussion in the Committee, although it resulted in a unique report on the use of human materials and information in biomedical research and development.[1]

2 OUTCOMES OF INITIAL DISCUSSIONS

As described, the resulting Kurokawa Report, which was produced as a formal and political report, was reasonable and included wide-ranging views on the issue. The report made recommendations on the standard of informed consent and the conduct of research ethics reviews by committees. However, it was not taken seriously by research communities or universities.

Back in the early 1980s, some medical schools were aware of the need to establish research ethics committees, since international medical journals required ethics approval as a mandatory process before the publication of a paper. Yet the Kurokawa Report did not go further to ensure that ethical review processes would be implemented. As a consequence, examples of unethical conduct have occured; a typical case is the article reported in *Science* about an incident of misconduct in biomedical research at Tokyo University.[2]

3 HUMAN SUBJECT RESEARCH IN JAPAN

In 2000 Prime Minster Keizo Obuchi initiated millennium projects for enhancing Japanese competiveness in science and technology (http://www.kantei.go.jp/jp/mille/). Substantial funding was made available for human genome research on causative traits for diseases. To begin with, the Ministry of Health and Welfare established a committee on human genome research projects. This resulted in the first guidelines on human subject research in Japan. They were finalised and published in April 2000.

The genome research guidelines were strongly associated with the funding scheme and awardees were required to carefully follow the guidelines. Research ethics committees were thereby established in almost all universities and research institutions. The guidelines were developed further in the

[1] http://www1.mhlw.go.jp/shingi/s9812/s1216-2_10.html, visited on 10 August.
[2] *Science* (2008), 321: 324.

following years, and we now have three major ethics guidelines on human subject research: for genomic, epidemiological, and clinical research.[3]

4 RESEARCHERS' CONDUCT UNDER THE ETHICS GUIDELINES

While I was observing research practices during this period of establishment of guidelines, I started to be involved in ethics committees as a review member. This experience made me realise that researchers do care a great deal about the guidelines, but they are not fully convinced. Of course, a function of the guidelines is to protect researchers by setting standard requirements, yet this function is not fully appreciated by the researchers or their communities.

The researchers and their communities are constantly complaining that the guidelines are unsuitable and impractical in their application. This is a natural consequence of the functioning guidelines, since they add complexity and inconvenience to research practices. Instead of simply complaining, it would be more useful if the researchers made suggestions for revising the guidelines in specific areas where they have issues with compliance.

5 JAPANESE DISCUSSION OF DATA PROTECTION LEGISLATION

From the end of the 1990s, the Japanese government initiated debate about the introduction of data protection legislation. This process took much longer than expected. As the Bill was being discussed, biomedical research communities, especially epidemiologists, together with newspapers and other mass media, actively appealed to be able to use personal data as part of freedom of expression in scientific research. This seemed to be successful and the Bill was amended to incorporate an exemption for academic use of personal data.[4] The exemption was granted in the legislation as enacted, but it also stated the need for self-regulation of the research community on the issue of data protection. However, the academic domain was satisfied

[3]http://www.mhlw.go.jp/general/seido/kousei/i-kenkyu/index.html.
[4]http://www.kantei.go.jp/jp/it/privacy/houseika/hourituan/.

with the exemption and did not move forward to discuss or establish self-regulation concerning the use of personal data in academic research.

6 WHAT DO THESE CASES MEAN FOR RESEARCH INTEGRITY?

I have presented the cases of establishment of research ethics guidelines and data protection legislation in Japan. In these cases research communities were very active until the establishment of the regulatory frames — they appeared to lose interest after the establishment of the regulations. However, the implementation of these regulations, especially in scientific research, requires self-control, since research needs to be carried out by individual researchers, even if working in a large team.

There is an interesting definition of science by Jonathan Rauch. It consists of these two phrases: no one has personal authority, and no one gets the final say. These phrases mean: it is totally contrary to science to say that "it is true, because I said so."[5] In other words, researchers are interested in continual, never-ending discussion. Of course in the real world, at some point we should stop discussing and start acting. But in this sense, what we do is never final and unchallengeable.

Another difficulty in the area of biomedical research is that biomedical science is more immediately concerned with clinical practice than other fields. This may need to be given close attention because there are essential differences between science and medicine. This point was made by R.H. Shryock in *Modern Medical History* (1937):

"Physicians were the only scientists who, because they were also practitioners of a vital art, were constantly being pushed to hasty and careless conclusions. Other research men, uncertain in the face of new problems, could suspend judgment and proceed with due caution. Practitioners confronted with dying patients did not dare to wait; they must act quickly and, if necessary, 'take chances.' Even during hours stolen for research, they were still under pressure to get practical results *as soon as possible*."

Given these thoughts, biomedical researchers and communities should develop their own independence, freedom, flexibility and responsibility when they conduct research using patients' materials and information.

[5] *Kindly Inquisitors: The New Attacks on Free Thought* (1993).

7 WHAT IS THE IDEAL SITUATION?

In the above cases, during the establishment of research ethics guidelines and personal data protection legislation, researchers appear to have become reactive and acquired characteristics of obedience and rigidity. These are the opposite qualities of what research integrity requires: freedom, flexibility and responsibility.

The Japanese legislation and the research ethics guidelines demand obedience, but it is very difficult to achieve simple obedience while still maintaining freedom, flexibility, and responsibility. We should therefore revise the roadmap of research integrity in Japan. We should reframe the functions of the external regulatory frameworks that affect research integrity.

Recently I encountered a shocking incident. A large Japanese biological research institute wished to establish a Brain Bank, and the researchers first asked legal scholars, to set the framework for their activity. This seems opposite to what I have in my mind. The researchers should first put all their effort into thinking, planning and developing an image of what they want to achieve. Only then should they consult legal scholars to make suggestions. It is essential that the process happens in this sequence, so that researchers can properly think about the scientific conduct of their research.

8 WHAT SHOULD HAPPEN NOW?

We need an informal exchange of opinion on what scientists need in practice and then move to a negotiation process about ethical requirements. Especially in Japan, researchers are socialised into obedience and seem to like to do as they are asked. It is the least economic and efficient way of doing science. The research communities seem to be afraid of the costs of being free, flexible and responsible. The incident at the University of Tokyo that I cited was considered as a kind of scientific misconduct, because researchers used unauthorised human materials in research. However, other important points should also be discussed, i.e. applicability of present guidelines on the legacy samples and the relationship between journal editors and researchers. These are the issues discussed in *Science*.[6] The concluding report was published by the Institute of Medical Sciences, University of Tokyo, and it was too reactive to the newspapers' views. It did not mention the scope of science

[6] *Science* (2008), 321: 474.

governance. It revealed precisely the sorts of discussion that could not be had on the bases of freedom, flexibility, and responsibility. The Institute was overwhelmed by mass media opinion. However, at that time, the scientific community should have taken the opportunity to show its integrity and engage in constructive discussion on the issue of the scientific misconduct. Because of its ingrained obedience, rigidity, and reactiveness, the scientific community of Japan could not achieve its mission. This should be remembered together with other incidents I cited in this report as efforts are made to establish research integrity in Japan.

CHAPTER 10

NATIONAL, INSTITUTIONAL AND INTERNATIONAL APPROACHES TO RESEARCH INTEGRITY: AN AUSTRALIAN PERSPECTIVE

Ren Yi

Research integrity management has become increasingly important, not only in Australia but also in the international arena. The main purpose of research integrity management is to reduce the risk while value-adding for the research community. Due to increasing competition for research funding and publications, the risk for research integrity management has increased dramatically, not only in developed countries but also developing countries. Cross-institutional research collaboration domestically and internationally has increased the complexity for research integrity management. All stakeholders such as governments, funding bodies, research institutions and researchers should be included through the management supply chain. This paper attempts to understand research integrity management at the national, institutional and international levels from the Australian perspective.

1 NATIONAL RESEARCH INTEGRITY MANAGEMENT STRUCTURE IN AUSTRALIA

Research integrity management can be summarised from both the policy and practice level. From the policy level, it focuses primarily on government policies. For example in Australia, the federal government, through its funding agencies such as the National Health and Medical Research Council (NHMRC) and the Australian Research Council (ARC), released the Australian Code for the Responsible Conduct of Research (ACRCR) in

2007 (NHMRC *et al.*, 2007). Similar codes also exist in the USA and UK. From a practice level, the policy usually involves funding research bodies, research institutions and researchers. Most of the research institutions in Australia have their own management policies on research misconduct, research ethics management, research management procedures and research financial management policies and procedures.

1.1 *Australian Code for the Responsible Conduct of Research 2007*

The Code has been jointly issued by the NHMRC, ARC and the universities of Australia. It is the fundamental guideline for conducting research in Australia. The purpose of the Code is to guide institutions and researchers to have responsible research practices. In describing good practice, the Code promotes integrity in research for researchers and explains what is expected of researchers by the community. In providing advice on how to manage departures from best practice, the Code assists researchers, administrators and the community in this important matter (NHMRC *et al.*, 2007, p. 1) (Australian Code for the Responsible Conduct of Research, 2007, p. 1).

There are two main parts of the Code. Part A primarily describes the principles and practices for encouraging responsible conduct of research for institutions and researchers. In other words, part A advises on what comprises good behaviour and good practice. A framework for resolving allegations of breaches of the Code and research misconduct has been provided in part B (NHRMC *et al.*, 2007). This also addresses the responsibilities from the institutional and individual levels.

1.2 *Australian Research Integrity Committee*

In April 2010, the Minister for the Department of Innovation, Industry, Sciences and Research (DIISR) announced the formation of the Australian Research Integrity Committee (ARIC) (DIISR, 2010). ARIC is an independent body which ensures institutions take appropriate action in response to allegations of research misconduct (DIISR, 2010). NHMRC and ARC will jointly manage ARIC as a national research integrity advisory body. ARIC will also review the processes followed by universities and other research institutions in response to allegations of research misconduct upon request. As the Australian research community is self-regulated, ARIC will supplement, rather than replace, existing institutional processes by reviewing the processes followed by institutions. ARIC is currently in the midst of being formed. It is expected that ARIC will be a fully functioning body by 2011.

2 INSTITUTIONAL LEADERSHIP IN ADVANCING RESEARCH INTEGRITY

Australian research institutions are taking great responsibility towards research integrity management. The majority of research institutions are publicly funded which therefore means they have a responsibility to ensure the use of public funding is done in a responsible and transparent manner.

The University of Southern Queensland (USQ) is a publicly funded research institution, located in Toowoomba, Queensland. USQ has more than 24,756 students and 1,987 staff (USQ, 2010a). Research strengths of USQ are Sustainability, including water management and sustainable business, Healthy Communities including rural and regional health, indigenous health, engineering innovation and flexible teaching and learning (USQ, 2010b).

USQ believes in providing a self-regulated environment. Institutional leadership is crucial in advancing and ensuring research integrity. Research integrity management has been categorised into Compliance, Misconduct management for research and Education for prevention at USQ.

2.1 *Compliance*

Compliance is an important part of research integrity management at USQ. It mainly involves administrative management, research management and financial management. Administrative management is mainly focused on the administrative roles and functions of compliance at USQ. The majority of research funding is from external funding bodies and therefore the funding guidelines and research contracts are basic compliance documents. The Code and institutional policies are other documents entered into by researchers. The university has a research management system in place to maintain the integrity of administrative management issues. Research management includes ethical approval for conducting research, raw data collection and storage, interpreting data and disseminating research findings through publications and other media. Financial management involves the responsible use of research grant funding which includes spending according to budget, no excessive abuse of financial systems, etc.

2.2 Research Misconduct Management

There are two research integrity management policies which have been updated and redrafted according to the 2007 code. The USQ policy on the Code of Responsible Conduct for Research and the USQ policy on Research Misconduct have been developed and updated based on past USQ policies as well as the 2007 code. The formality of policy making is complicated at USQ. Three offices, the Office of Research and Higher Degrees (ORHD), the Human Resource Office (HR) and the Legal Office (Legal), have been involved with the initial drafting of the two policies and a working party has been established between the three offices. The USQ policy on the Code of Responsible Conduct for Research includes:

- Principles of managing research integrity at USQ
- University, faculty and researcher's responsibilities
- Research data management
- Confidentiality
- Ethical considerations
- Authorship
- Research supervision
- Conflict of interests (USQ, 2010c)

The USQ policy on Research Misconduct includes:

- Definition of misconduct in research
- Preliminary investigation
- Formation of special investigation committee (USQ, 2010d)

Two drafted policies have received university-wide consultation, all faculties and centres have been informed and consulted through the university research management structure. All feedback has been included in the next draft approval sought from the University Research Committee. A majority of members on the committee have contributed during the consultation phase and the policies have been approved and endorsed by the committee. Draft policies have been submitted to the Union through the Human Resource office. Research misconduct at USQ has been categorised as an industrial relations matter and the Office of Research and Higher Degrees collated and inserted all comments from union and university research committee in the draft paper, which was then sent to the Vice Chancellor's Committee (VCC) for approval. In the meantime, the policies were sent to an academic board for noting. The Director, ORHD presented the policies at VCC. Following only minor comments by members of VCC, the draft policies were endorsed and sent to the University Council for final approval. The policies become effective from the date they are approved by the University Council.

2.3 *Education for Prevention*

Education is key for research integrity management at USQ. Different programs have been adopted for researchers and higher degree research students (HDR students) at USQ. The programs include:

- New staff induction program
- Research integrity workshops
- HDR student orientation program
- HDR student handbook

The main purpose of these programs is to disseminate the National Code of Conduct for Responsible Research as well as university policies. Awareness of these documents is important, particularly for staff and HDR students from overseas who need to adopt and embrace the Australian and USQ research environment. These programs are value adding for research integrity management at USQ and reduce the risk for the university around research misconduct cases.

3 RESEARCH INTEGRITY MANAGEMENT FOR INTERNATIONAL RESEARCH COLLABORATION

Research integrity management for international research collaboration is a challenging issue these days. There are no frameworks or guidelines in place around international research collaboration. There are no international bodies or government agencies that are willing to take responsibility for the management of this issue. There has only been one publication to date which addresses this issue. It is an OECD publication entitled *Investigating Research Misconduct Allegations in International Collaborative Research Projects: A Practical Guide* (OECD, 2010). Meanwhile, more and more international research misconduct allegations have been investigated from many countries and institutions.

International science and research communities, including governments, funding bodies, research institutions and researchers around the world are agreed that international research collaboration is essential for innovation of systems in the knowledge economies. Therefore, an international research integrity management framework is crucial for governing and guiding the international research community. The proposed international research integrity management framework should be embedded in an international organisation such as the United Nations (UN) or the United Nations Education, Scientific and Cultural Organisation (UNESCO). The international research community should also have agreed principles on research integrity management. The Singapore Statement from the 2nd World Conference on Research Integrity could be the base for the international research community, particularly governments, adopted as agreed principles on research integrity management (WRIC, 2010).

4 CONCLUSION

This paper has argued the national, institutional and international approaches to research integrity from an Australian perspective. The Australian research integrity management framework has been identified and highlighted in the paper. Responsibilities for research integrity management from national, institutional and individual levels have been explained. The University of Southern Queensland has been used as a case study for advancing institutional research integrity management. For international research integrity management, an agreed principle, a framework and a responsible international organisation have been proposed for the international research community.

References

1. Australian Research Council (2010). *Australian Research Council (ARC) Fact Sheet*, www.innovation.gov.au/Section/AboutDIISR/FactSheets/Pages/AustralianResearchCouncil%28ARC%29FactSheet.aspx. Accessed on 18 August 2010.
2. Department of Innovation, Industry, Sciences and Research (DIISR) (2010). *Australian Research Integrity Committee (ARIC) Fact Sheet*, http://www.innovation.gov.au/Section/AboutDIISR/FactSheets/Pages/AustralianResearchIntegrityCommittee%28ARIC%29.aspx. Accessed on 11 October 2010.
3. National Health and Medical Research Council (2009). *Current Outcomes of Project Grants Funding Rounds*, www.nhmrc.

gov.au/grants/rounds/projects/index.htm. Accessed on 18 August 2010.

4. National Health and Medical Research Council., the Australian Research Council., & Universities Australia (2007). *Revision of the Joint NHMRC/AVCC Statement and Guidelines on Research Practice: Australian Code for the Responsible Conduct of Research.* Australian Government: Canberra, http://www.nhmrc.gov.au/publications/synopses/r39syn. htm. Accessed on 18 August 2010.
5. Organisation for Economic Co-operation and Development (OECD) (2010). *Research Integrity: Preventing Misconduct and Dealing with Allegations*, http://www.oecd.org/document/13/0,3343,en_2649_34319_42713613_1_1_1_37417,00.html. Accessed on 13 October 2010.
6. Organisation for Economic Co-operation and Development (OECD) (2009). *Investigating Research Misconduct Allegations in International Collaborative Research Project: A Practical Guide.* OECD Global Science Forum, http://www.oecd.org/dataoecd/42/34/42770261.pdf. Accessed on 12 October 2010.
7. University of Southern Queensland (USQ) (2010a). *USQ Facts and Figures*, http://www.usq.edu.au/aboutusq/facts. Accessed on 14 October 2010.
8. University of Southern Queensland (USQ) (2010b). *USQ Research Action Plan*, http://www.usq.edu.au/research/usqresearch. Accessed on 20 May 2010.
9. University of Southern Queensland (USQ) (2010c). *Code of Responsible Conduct for Research*, http://policy.usq.edu.au/portal/custom/detail/research-and-scholarship-policy/index.html. Accessed on 14 October 2010.
10. University of Southern Queensland (USQ) (2010d). *Research Misconduct*, http://policy.usq.edu.au/portal/custom/detail/research-and-scholarship-policy/index.html. Accessed on 14 October 2010.
11. World Conference on Research Integrity (WCRI) (2010). *Singapore on Research Integrity*, http://www.singaporerestatement.org. Accessed on 12 October.

CHAPTER 11

FINLAND: HOW TO REVISE NATIONAL RESEARCH INTEGRITY GUIDELINES IN THE CHANGING INTERNATIONAL LANDSCAPE?

Eero Vuorio

The history of research integrity in Finland goes back to November 1991 when the National Advisory Board on Research Ethics was established by a government decree. The statutory task of the Board is to promote discussion on and to disseminate information about research ethics in Finland and to take initiatives concerning research ethics.

1 RESEARCH ETHICS *VERSUS* RESEARCH INTEGRITY — TRANSLATION OF TERMS INTO NATIONAL LANGUAGES

The term Research Integrity is difficult to translate into some languages where the term integrity is essentially restricted to mean "personal integrity". Therefore in several European languages Research Ethics rather than Research Integrity is used in the names of boards and guidelines dealing with Research Integrity. This may appear confusing since other ethics boards are in place for *ethical evaluation* of research proposals dealing with humans, experimental animals, gene technology etc. In the Finnish Guidelines, the term "good scientific practice" is used in the same meaning as "responsible conduct of research". Violations or deviations of good scientific practice are divided into two categories, misconduct and fraud. Although the terminology used in different countries may appear divergent, there seems to be very good agreement on how scientists should behave, i.e. what is good scientific practice or responsible conduct of research.

2 FINNISH GUIDELINES

The most visible activity of the Advisory Board has been drafting of national research integrity guidelines: *Good scientific practice and procedures for handling misconduct and fraud in science* (http://www.tenk.fi/ENG/Publicationsguidelines/htkeng.pdf). Drafted together with members of the scientific community, the guidelines have received wide acceptance: all Universities and Polytechnics, a great majority of other research-performing organisations, major funding organisations and several learned societies have agreed to adhere to the guidelines and the procedures for investigating alleged violations. This is important since the effectiveness of the guidelines depends on the commitment of the research community to comply with them.

The first part of the guidelines discusses the principles of the concept of good scientific practice. A lot of emphasis is given to the fact that commitment to good scientific practice is not only a duty of each researcher and each member of a research team, but also a responsibility of the research community as whole. The second part of the guidelines describes different types of violations or deviations from good scientific practice, which are divided into two categories: misconduct and fraud. Manifestations of fraud are grouped in four categories: fabrication, misrepresentation, plagiarism and misappropriation. The third part of the guidelines contains a detailed procedure for handling alleged violations of both types. This includes a written notification to the rector/director of the institution where the suspected violation has occurred, an inquiry, an investigation, a final report and a decision by the rector/director on possible sanctions. Depending of the severity and possible recurrence of misconduct the sanctions may range from an oral or written warning to failure to approve a thesis or obtain a degree, to more severe punishments as set in the Universities Act (and other legislation). Special emphasis is made to retract or correct any publication containing information judged to contain fraudulent material, and to publish the findings of the investigation in the same forum. If any of the parties concerned are dissatisfied with the procedure as carried out at the local level, they may write to the Advisory Board for an opinion. The Advisory Board receives copies of all cases investigated locally for its files and statistics.

3 REVIEW OF THE CURRENT SYSTEM

After 18 years of operation, one is probably justified to make some conclusions based on experience. In principle, the system functions quite well, but an obvious problem is the small number of investigations of alleged misconduct or fraud cases that rectors of universities and polytechnics or directors of research institutions have to conduct during their term. Although they are aware of the Guidelines, they have little experience in carrying out the procedure. Up until recently, universities had no designated personnel for practical handling of inquiries and investigations. This probably explains why some inquiries and investigations carried out exhibited slight deviations from the Guidelines, particularly in respect to the timelines. Today, many institutions have permanent Research Integrity Boards composed of members of the scientific community that help in investigating alleged misconduct or fraud at local level.

Despite continuous training efforts, some researchers do not seem to be aware of the concept of responsible conduct of research or of the Guidelines. All doctoral training programs now contain some lectures on Research Integrity and the doctoral trainees involved are well aware of the Guidelines. However, some researchers who carry out their doctoral studies outside formal training programs and some international visiting scientists seem to be completely unaware of them. The number of misconduct cases involving foreign scientists appears to be increasing, possibly reflecting the overall increase in the international mobility of researchers. Lack of approved international guidelines is also seen as a problem. According to the Finnish Guidelines all allegations should be investigated in the laboratory where the work under suspicion was carried out.

4 REQUESTED IMPROVEMENTS

As Finnish Guidelines have been drafted for the scientists by representatives of the scientific community, they have also the possibility to amend them. Wide recognition of the Guidelines throughout the research community has, however, increased the threshold for revising them. Revised Guidelines would obviously require a new round of consultation of the nearly 100 institutions currently following them and a massive training effort. Among the key revisions requested are extension of the Guidelines to cover issues such as falsifications in self-drafted *curricula vitae* and publication lists, in non-scientific communication with media, and expert statements by researchers to courts, public decision makers etc. The current Guidelines only cover scientific communication. As the aim of the Advisory Board is to maintain the Guidelines consistent with any internationally agreed guidelines, any revision of the current 2002 Guidelines has been delayed until such international guidelines are available.

CHAPTER 12

ACTING AFTER LEARNING IN EUROPE

Dirk G de Hen

From an European point of view it is good to see so many participants present from different parts of the world. And we are not here to talk about science fiction, but we are most interested in the welfare of good, reliable science and research. Although there seem to be countries without threats to such research, I may boldly suspect that this is because in those countries the main stakeholders refuse to do research into those threats or, having done research, refuse to acknowledge the existence of threats. We are gathered here to talk about means to avoid and/or battle against these threats and in particular to preserve research integrity.

Preserving research integrity begins with the recognition that research is carried out by humans, each human with his or her own character, education and personal weaknesses. People and thus researchers-to-be can learn and so each student should learn how to behave well when they want to become respected researchers, whose work will contribute to scientific progress.

In parts of Europe, serious thinking about research integrity started some 30 years ago, and the awakening has not yet been completed. Nevertheless serious attempts have been made and successes have been reached, mostly beginning at some individual institutions and set up by individuals. We have seen dozens of codes of conduct or good practice, locally, nationally and lately internationally. I refer here to a Code of Conduct, made by the OECD, and a European Code of Conduct by the European Science Foundation.

Like in other parts of the world, people in Europe tend to know regulations in general. However, this does not happen to be the same as following the rules. When this situation occurs in relation to research integrity, one can see that in many cases individuals started to make complaints about wrong behaviour. In a lot of cases they wanted to make a complaint not for the sake of good research in general but because they felt harm was done to their own work or to their own scientific or research reputation.

Compared to 30 years ago, both national and international developments have given the scientific world and the research world, including researchers in industries, a number of both codes as well as procedures. Due to differences in the legal systems and provisions and a multitude jurisprudence between countries, it seems that until now no single solution with one legally binding text can be produced for those counsellors, committees and eventually national research integrity offices who seek assurance in internationally approved texts.

Nevertheless I feel this may become a possibility in the future as the world has also produced texts and procedures for the International Court of Justice and for international arbitration. In the meantime, it is my strong belief that we can learn a lot by taking action and exchanging information.

For this goal, ENRIO, the European Network of Research Integrity Offices, was established in 2007. It consists of participants from 14 European nations and this number may increase by next year to probably 16 or 17.

ENRIO's aims are:

- To facilitate discussions and share experiences and possible solutions related to the investigation of allegations of misconduct in research and related to the training and education with regard to good practices for research;
- Secondly to report discussions and develop proposals for submission to national and international organisations on matters of investigations and training; and
- To liaise and work in partnership with other organisations with European or global interests in research integrity.

The experience within Europe is rather dissimilar. There are countries with very little experience in the field of investigations of allegations of misconduct or in the field of training and education about research integrity, and there are countries with more than 25 years of experience with a well-developed system on both local and national levels. For both "types" of countries, learning from other experiences is well accepted and well appreciated. I strongly believe that this learning from others may well contribute to the knowledge of all parties concerned. It may help in any country not only to be able to compare its own situation with the situation elsewhere, but also proceed with success in institutionalising research integrity at the own national level.

One important remark I would like to end with is that the basis for success of ENRIO is also the fact that the cooperation between its members is fully voluntary and without bureaucracy. The main obligation is to contribute to the goals as much as possible and thus to help encourage researchers to act according to the rules of good practice and to encourage institutions to be open about cases of misconduct, rather than being weak and hunted down for the fact that misconduct happened at their institution. Researchers should be open because they are proud to let the world know that they strive for the increasing reliability of science, research and the people working in it.

CHAPTER 13

VIEWS ON RESEARCH INTEGRITY IN THE COMMONWEALTH OF INDEPENDENT STATES

Boris Yudin

This paper discusses an attempt to cope with research misconduct in Russia as well as in countries of the Commonwealth of Independent States (CIS). CIS includes most of the post-Soviet states. These countries have a lot of common traits, such as cultural traditions (including research culture), wide proliferation of the Russian language etc.[1]

In 2010, we had started a project which is sponsored by UNESCO's Moscow Bureau. The project, which is headed by Professor Ruben Apressyan (Moscow), aimed at the elaboration of the Ethical Code of Scientists for our countries. At the first stage of the project, a questionnaire was prepared and distributed among researchers and science managers in CIS countries. The questionnaire includes such items as:

Do you think that the following misuses must be pointed out in the Code:

- plagiarism;
- false co-authorship;
- fabrication of data;
- falsification of data?

[1]See, for example: Kubar *et al.* (2007). *Ethical & Review of Biomedical Research in CIS Countries (Social and Cultural Aspects).* Saints-Petersburg, 2007.

Who must develop standards of responsible conduct of research:

- scientific community;
- experts;
- professional scientific associations;
- national academies of sciences;
- appropriate ministries and agencies;
- international scientific organisations?

Who must monitor the compliance with standards of responsible conduct of research:

- scientists;
- administrative authorities of the National Academy of Sciences;
- ethical committee of the National Academy of Sciences;
- scientific council of the university or research institute;
- ethical committee of the university or research institute;
- administrative authority of the university or research institute;
- authorized departments of appropriate ministries and agencies;
- professional scientific associations?

Who must inspect the quality of research results:

- scientists;
- administrative authority of the university or research institute;
- ethical committee of the National Academy of Sciences;
- scientific council of the university or research institute;
- ethical committee of the university or research institute;
- mass-media, journalists;
- authorised departments of appropriate ministries and agencies;
- professional scientific associations?

Who must perform an inquiry after a claim of research misconduct has been made:

- scientists;
- mass-media, journalists;
- investigating authority, office of prosecutor;
- professional scientific associations;
- administrative authority of the university or research institute;
- authorised departments of appropriate ministries and agencies;

- ethical committee of the university or research institute;
- scientific council of the university or research institute?

There were also such questions as:

- Who must perform an inquiry when a claim on plagiarism is made?
- Which sanctions should be imposed in the case of research misconduct?
- What is the goal of an inquiry on research misconduct?
- To whom should the Code must be addressed?

It is evident that in some of these questions respondents are asked to choose between the involvement of scientists or, generally speaking, research communities, on one side, and of some external authorities, on the other side, in different activities related with research misconduct. For the time being we have received only a small number of filled-in questionnaires, so that we cannot make even minimally justified conclusions. Nevertheless, these rather limited data are indicative of one essential tendency. All our respondents turned out to be rather suspicious with regard to possible participation of persons or authorities outside the research community in making policies and decisions on research conduct/misconduct.

Of course, this tendency can be revealed not just in the CIS countries, but in any part of the world. It corresponds to the traditional ethos of science: science is seen as an autonomous activity, performed by the scientific community. Autonomy in the classical sense of the word means that it is the community that develops norms of its own activities and monitors fulfillment of these norms by its own members. So, every outside intervention is perceived by the scientific community as a threat to its autonomy.

In the case of the CIS and other post-Soviet countries, this tendency also has some specific historical roots: science in the Soviet Union had suffered from the strong ideological and political pressure exerted by the Communist Party and the State. Most scientists in post-Soviet countries are aware of many severe perversions, which took place in the 1940s and 1950s in different fields of science, including genetics, cybernetics, quantum chemistry etc. That made them rather sensitive to any signs of outside intrusion into research.

However, this is far from the full story. The problem of research misconduct has not only historical, but contemporary roots as well. Nowadays researchers are responsible not only to the research community, but to many other different stakeholders, often with rather divergent interests. In some cases we can speak about the vested interests of some stakeholders in ensuring fudged research results. These vested interests are among the factors that make the problems of research integrity so acute.

We can distinguish between **first order**, i.e. more immediate **responsibilities of a researcher** — to the scientific community and to sponsor and his/her **second order responsibilities** — to the state (if a research project is not supported by the state agencies) and to society (more exactly, to different social groups which are interested in the results of any given research). Strictly speaking, among all stakeholders only the scientific community is **institutionally** (i.e. not only pragmatically, instrumentally) interested in research integrity.

Of course, we can and must speak about the loose organization of the scientific community as a rather specific social system. The system also includes other institutional arrangements (such as competition inside the community), which often impede research integrity and negative effects of which need to be taken into account and neutralised as much as possible. Nevertheless, any efforts to cope for research integrity apart from the scientific community would be counterproductive. Only those measures which will be approved (at least tacitly) by the community can be really effective.

SECTION III

RESEARCH MISCONDUCT

INTRODUCTION

As countries increase their emphasis on research, the need to conduct investigations of reported problems is likely to grow. This requires clear definitions of unacceptable behaviour, known more commonly as "research misconduct", a task that is not as simple as one might expect. The opening paper in this Section by Fanelli illustrates this point with a description of the legal and scientific battles in the US over a definition. In the end, in 2000, the US Office of Science and Technology Policy adopted a narrow, legalistic definition focused on three sets of behaviour: "fabrication, falsification and plagiarism" (FFP), ignoring other questionable behaviour that was covered in previous policies. Other countries have adopted similar definitions, particularly those whose policies are based on quasi-judicial structures. Where differences in policy have emerged, it is often over the issues of "intentional" vs. "non-intentional" acts. The Lomborg case in Denmark has helped crystallise thinking about the intentionality of misconduct and questionable practice. In Denmark, this resulted in the addition of "undue misrepresentation of a person's scientific work and/or scientific results" in the definition of misconduct. Fanelli closes his analysis with a call for broader definitions, to include not only the presumed "black and white" FFP but also other grey areas that can undermine both the reliability of the research record and trust in research.

Vaux, in the Conference Keynote Address, took as his theme, two aspects of integrity — the integrity of the research record and the integrity of the individual researcher. With regard to the research record, Vaux contends that there has been an alarming growth in publication bias towards positive results, especially in the life sciences. This growth is a matter of concern that needs to be addressed by the research community and the scientific publishing world. Vaux also offers several examples that suggest that the integrity of the research record is also adversely impacted by carelessness or incompetence. This should be considered separately from the numerous cases of misconduct. Vaux stresses that these shortcomings, however they originate, need to be corrected. Detection in these cases requires the vigilance of readers/researchers to identify errors so that they can be corrected. Vaux feels that it is the duty of the readers/researchers to take appropriate action to correct the record. This has implications for the individual researcher and her/his responsibilities.

Martin, in his fascinating account of a serial plagiarist, questions whether self-policing really works. This tale of plagiarism illustrates five points: (1) research misconduct occurs in many disciplines, in this case, the social sciences, (2) plagiarism is not a "victimless crime", a stance taken by some people. There are many who can claim to be victims from the events described, (3) self-policing can totally break down due to the inadequacies in the peer review system. This has also been shown in other cases, most notably the Schoen case at the Bell Laboratories, (4) even when self-policing works, it can take many years for this to occur, thus wasting precious time and resources, and (5) sometimes, those who embark on or drift into misconduct find it very difficult to stop. As the proverb says, "one lie may lead to another."

The final contributions in this section, by Kruglyakov and Cuellar respectively, address profound problems of political structures in their own countries (Russia and the former Soviet Union in one case and Mexico in the other) in relation to research integrity. The message is clear. Research can only flourish within open government structures and when a culture of transparency prevails. As we can probably all attest, these two countries are not the only ones given to secrecy. It demonstrates, especially in the Russian example, that there is a need, when enforcing research integrity, for scientifically literate political elites coupled with transparent and independent research processes, especially in relation to funding.

Together, these papers call for common and broad standards of research responsibility that cover not only individuals and research institutions but extend more widely to academic and political governance, if researchers are to win and maintain public trust in and support for research worldwide.

CHAPTER 14

THE BLACK, THE WHITE AND THE GREY AREAS: TOWARDS AN INTERNATIONAL AND INTERDISCIPLINARY DEFINITION OF SCIENTIFIC MISCONDUCT

Daniele Fanelli

A global consensus on what constitutes good and bad research practice would seem easy to achieve — after all, if the laws of nature do not change across nations or fields of research, why should the criteria to investigate them change? In practice, however, this objective has proven to be elusive and complex. The definition of research misconduct, in particular, has been the subject of heated debates and frequent revisions in most countries that have adopted one. Reflecting on such debates, some scholars have doubted whether an agreement on such matter is at all possible, and worth the effort [1]; others, however, have simply proposed new approaches to overcome the obstacles [2, 3]. Although scepticism is sometimes expressed even in international reports (e.g. [4]), all recent global initiatives, including this Second World Conference on Research Integrity, call for at least an harmonisation of principles and/or for agreements between international collaborators at the start of each research project, [5–7].

Why is unification desirable? Research is increasingly international and interdisciplinary [3–5]. Yet, institutions in many countries either have different definitions or lack them altogether [3, 7–9], and scholars in different disciplines might have different standards of research practice [10] or even hold different beliefs about objectivity and the possibility to distinguish facts from values [11]. These latter differences in beliefs are a matter of philosophical debate, which can only be resolved through academic research. The other differences, however, are not, and smoothing them out would bring several advantages. It would facilitate investigations in international contexts, including when researchers accused of misconduct relocate to a new country [12]. It could be used as a template for fields and countries that currently lack any official definitions [3]. It would provide a universal standard for education of the responsible conduct of research, which would harmonise the quality of scientific results and increase the trust between researchers in different countries and communities [7]. It would benefit scientific research on the causes and remedies of misconduct itself, by bringing greater clarity and consistency of concepts [7].

How far are we from reaching an agreement? What are the practical obstacles that we currently face? In this paper, I will first overview how research misconduct has been defined, over the years, in different countries. This will help to illustrate the essential terms of the debate. I will then compare some of the most recent definitions given in different countries, to illustrate the current diversity, discuss its causes and suggest possible ways forward.

1 THE DEBATE IN THE US AND OTHER COUNTRIES

The contemporary debate on defining research misconduct famously ignited in the early 1980s in the USA. Following a series of scandals and political initiatives, in 1986, the Public Health Service published its first official definition of misconduct: "(1) serious deviation, such as fabrication, falsification or plagiarism, from accepted practices; (2) material failure to comply with federal requirements affecting specific aspects of the conduct of research (e.g. the protection of human subjects and animal welfare in labs)" [13].

The above definition was criticised for being too vague and for not making a clear distinction between intentional and unintentional acts. Following a series of public hearings and discussions, the PHS redefined misconduct in 1989 as: "fabrication, falsification, plagiarism (FFP), or other practices that seriously deviate from those that are commonly accepted within the scientific community for proposing, conducting, or reporting research. It does not include honest error or honest differences in interpretations or judgements of data." This definition solved the problem of intentionality by excluding honest mistakes, but was only formally less vague than the previous one. Although it set a standard that many countries are still adopting today (e.g. Norway, as explained below), in the USA it was criticised for at least three reasons. First, it does not specify the behaviours that are unacceptable, thus infringing the fundamental legal right to know in advance those activities that are proscribed [14]. Second, it could potentially include, and therefore hamper, the most pioneering and creative research, which by definition goes beyond what are "commonly accepted" practices [13–15]. Third, it opened the doors to virtually all forms of misbehaviour, including those not strictly related to research. This definition was the basis, for example, for an allegation of sexual harassment, which critics considered a "preposterous and appalling application of the definition" [14].

A new standard in the US was set in 1992 by a joint panel by the National Academy of Sciences (NAS), the Institute of Medicine and the National Academy of Engineering, who in a consensus report introduced important novelties. Misconduct was limited to "fabrication, falsification, or plagiarism, in proposing, performing, or reporting research", with falsification meaning "changing data or results". Any other misbehaviour fell into the category of "Questionable Research Practices" (henceforth indicated as QRP, as opposed to FFP), for which there was no consensus "as to the seriousness of these actions nor [...] on standards for behaviour in such matters". This category included behaviours such as guest authorship, withholding research information, and misuse of statistics to enhance the significance of research findings [15]. Although other, completely different approaches to defining misconduct were later proposed (e.g. "misrepresentation, misappropriation and interference" by a Commission on Research Integrity in 1995 [16]), the current US federal definition matches the NAS 1992 narrow "FFP" definition, and makes no mention of its QRP category. Its definition of falsification, however, includes a wider range of misbehaviours (i.e. "manipulating research materials, equipment, or processes, or changing or omitting data or results such that the research is not accurately represented in the research record") [17].

The US federal definition of misconduct might be considered the most narrow, legalistic and clear-cut definition adopted by any country. Yet, its clarity and precision are more apparent than real. A finding of misconduct in the US requires evidence of "a significant departure from accepted practices of the relevant research community" and that "the misconduct be committed intentionally, knowingly, or recklessly" [17]. Therefore, the old problem of defining consistently and objectively what a "serious" deviation entails has not gone away. Nor have other problems, for example in deciding when one can be held responsible for others' actions. The many ambiguities and confusion that this definition creates, both scientifically and legally, are continually being discussed [18].

Definitions of misconduct in some countries have undergone, over the years, a similar narrowing to that occurred in the USA. In Norway, for example, the National Commission for the Investigation of Scientific Misconduct had, in 1994, a very broad definition: "all serious deviation from accepted ethical research practice in proposing, performing, and reporting research" [19]. Thirteen years later, with the establishing of a new body, the National Commission for the Investigation of Scientific Misconduct, the definition was revised and aligned to USA's 1989 standard, i.e . "falsification, fabrication, plagiarism and other serious breaches of good scientific practice that have been committed wilfully or through gross negligence when planning, carrying out or reporting on research [20].

In other countries, however, the debate evolved in the opposite direction, with the definition of misconduct being broadened. In Finland, for example, the National Research Ethics Council, in 1994, was concerned with "the presentation to the scientific community of fabricated, falsified or misappropriated observations or results, for example in a presentation held in a scientific meeting, a manuscript written for the publication or a research-grant application [...]" [19]. In 2002, the definition was revised and became much broader and articulated (too long, in fact, to be accurately quoted here). It includes two categories: "fraud" and "misconduct". The former corresponds to FFP, the latter is even broader than USA's QRP category, because it includes forms of carelessness and negligence (i.e. "negligence in referring to earlier findings", "careless, and hence misleading, reporting of research findings and the methods used", "negligence in recording and preserving results", etc.) [21]. Indeed, the Finnish definition only explicitly excludes "honest differences in interpretations or judgments of data" and says nothing to exclude honest errors, setting the minimum level of intentionality very low.

Ultimately, it seems that the "evolutionary history" of misconduct definitions has been different in every country. As a final example, we can take Denmark, one of the first European countries to establish a national body dedicated to misconduct investigations. The Danish Committees on Scientific Dishonesty, in 1992, defined misconduct as "intention or gross negligence leading to falsification or distortion of the scientific message or a false credit or emphasis given to a scientist" [19]. This definition was expanded in 1998 to "actions or omissions in research which give rise to falsification or distortion of the scientific message or gross misrepresentation of a person's involvement in the research", including "consciously distorted reproduction of others' results", and "consciously distorted interpretation of results and distortion of conclusions" [22]. On the basis of this broad definition, in January 2003 the DCSD judged the political scientist Bjørn Lomborg guilty of "objective" scientific misconduct ("objective" meaning that the intention to deceive was not proven) because of alleged bias and imprecision in his controversial book "The Skeptical Environmentalist". This ruling attracted severe criticisms from parts of the academic community, and called for a reexamination of DCSD's criteria and procedures [23, 24]. In the meantime, inspired by the Lomborg case, Danish academics started to use misconduct allegations as a rhetorical weapon against controversial research done by colleagues [25].

In clear analogy to the US debate, the working group set up by the Danish Research Agency criticised the DCSD's definition because it combined intentional and possibly non-intentional acts, and because it used vague terms and concepts like "distortion" and "scientific message". It was also emphasised how even apparently clear-cut concepts like "fabrication and construction of data" cannot be applied with the same criteria across all disciplines [26]. The Danish definition of scientific dishonesty thus returned to a more standard, but still relatively broad "intentional or grossly negligent conduct in the form of falsification, plagiarism, non-disclosure or any similar conduct involving undue misrepresentation of a person's own scientific work and/or scientific results", with interpretative biases being considered misconduct only when "undisclosed" [22]. A new amendment to the law has endorsed an even more concise and standard definition: "falsification, fabrication, plagiarism and other serious breaches of good scientific practice that have been committed wilfully or through gross negligence when planning, carrying out or reporting on research", although this is then followed by the usual, comprehensive list of misbehaviours [27].

2 THE BLACK, THE WHITE AND THE GREY AREAS

Even though definitions have evolved in different ways in each country, the debates underlying these changes have taken place along the same two main lines of contention. One line is concerned with how many behaviours the definition should include. At one extreme, we have the very narrow FFP approach, which is supposed to coincide with undeniably "black" areas of scientific practice. Many definitions, however, are conceptually open to include any potential breach of integrity, thereby embracing the "grey areas" of research conduct. Whether open or closed, definitions might also include unethical behaviours not strictly linked to research practices, thus entering "white areas" of generally unethical behaviour, which few people would contest but many might wish to keep separate from the issue of research misconduct. The second line of contention is concerned with the minimum level of intentionality — the equivalent of what in criminal law is the *mens rea* (Latin for "guilty mind") of the offender. Although legal systems vary considerably in the levels of culpability and the criteria to distinguish them, in the context of research misconduct three basic categories can be recognised: intentional acts (i.e. done with the intention to deceive), grossly negligent acts (i.e. done in "reckless disregard for the truth" [16]), and careless or negligent acts (i.e. which discard "the standards of a reasonable, normal person" [25]).

If we now look in detail at some of the definitions currently adopted in different countries by various institutions, we see some common features but also a considerable diversity, both in broadness and in level of intentionality (Tables 1 and 2, respectively, which indicate presence/absence of specified elements within each definition). All definitions share a "core" of misbehaviours corresponding explicitly or implicitly to FFP, but then usually go beyond this by having an "open" definition and/or specifying additional misbehaviours of various kinds (Table 1). Similarly, all definitions presumably include wilful acts of misconduct (although they do not necessarily specify this), but most go beyond and explicitly include or exclude other levels of intentionality, in different combinations (Table 2).

Where does this diversity come from? Ethical analyses agree that part of the explanation lies in different goals that the definitions are set to achieve. When the goal is to hold researchers accountable for fraud, then a narrow, closed definition is preferred. If, however, the goal is to promote responsible

Table 1

country	year	institution	fabrication and/or falsification and plagiarism	open definition	selective reporting	ghost-guest authorship	misusing statistics	misrepresenting others' research	sabotaging others' research	biased interpretation of results	mismanaging conflicts of interest	duplicating publications	not following pre-approved protocols	mismanaging or not preserving data	misrepresenting professional credentials	favouring misc.- hampering misc.- investigations	abusing peer reviewer power	withholding information or materials from colleagues	financial misconduct	personal abuse	bad mentorship	harming human or animal subjects	other	source
AU	2007	NHMRC et al.	x	x		x					x		x			x			x			x		[29]
CN	2009	CAS	x	x		x			x			x						x	x				x	[30]
CR	2007	CESHE	x		x	x		x	x	x	x	x											x	[31]
DK	2009	DCSD	x	x	x		x			x													x	[27]
FI	2002	TENK	x	x	x		x	x	x	x		x		x									x	[21]
FR	1999	INSERM	x			x					x			x							x		x	[32]
IN	2006	ICMR	x		x	x		x			x	x					x						x	[33]
NL	2001	KNAW et al.	x	x	x	x	x	x		x			x		x					x			x	[34]
NO	2007	NCISM	x	x																				[35]
SW	2004	EGISRM	x	x	x		x																x	[36]
CH	2003	SAAS	x	x	x	x	x	x	x	x				x		x	x	x					x	[37]
UK	2009	UKRIO	x	x		x							x											[38]
US	2005	PHS	x		x																			[17]

Table 2

country	year	institution	intentional	grossly negligent or reckless	negligent	excludes honest errors	excludes differences in opinion	source
AU	2007	NHMRC et al.	x	x	x	x1	x	[29]
CN	2009	CAS				x	x	[30]
CR	2007	CESHE	x			x	x	[31]
DK	2009	DCSD	x	x				[27]
FI	2002	TENK		x	x		x	[21]
FR	1999	INSERM	x	x	x		x	[32]
IN	2006	ICMR						[33]
NL	2001	KNAW et al.	x		x			[34]
NO	2007	NCISM	x	x				[35]
SW	2004	EGISRM			x			[36]
CH	2003	SAAS	x		x			[37]
UK	2009	UKRIO						[38]
US	2005	PHS	x	x		x	x	[17]

research, foster research integrity, or protect the scientific literature from false and biased findings, definitions must enter the grey areas, and therefore should be broader and have lower intentionality thresholds. Moreover, if the aim is to promote higher ethical or political values, then definitions will go beyond research practice and enter the "white" realm of generic ethical values [27–29]. Given the diversity of national approaches and structures currently in place to address misconduct (i.e. some have national bodies with investigative powers, others have national bodies with purely advisory roles, others have only institutional offices, etc.), it seems logical to expect a parallel diversity of definitions. However, this cannot fully explain national differences, because countries with similar systems (e.g. Denmark and the USA, both with national investigative bodies) have rather contrasting definitions.

Another shaping force in misconduct definitions are the underlying beliefs about the capacity of research communities to self-correct and self-police, and beliefs about the level of control that society should exert on research [2]. The latter are obvious in the "white area" type of definitions, which explicitly aim to keep science within predetermined ethical, social or even political borders. Narrow definitions, on the other hand, express a strong belief in academic independence and self-regulation. Indeed, as we have seen, one of the main criticisms moved against "open" definitions is that they might discourage pioneering research. Although there is no recorded case of novel research being treated as misconduct in the USA [16, 28], the Danish experience suggests that such risks do exist. The Lomborg case in Denmark, moreover, revealed how the criteria for "good" research practice and the mechanisms of self-correction and self-policing can differ dramatically between disciplines.

A third, and perhaps neglected, factor that might complicate matters further is purely linguistic. In some cases, different words are used to express the same concept. This is typically, but not exclusively, a problem of translation. The English language, for example, clearly distinguishes between "ethics" and "integrity" [1], but the Finnish language lacks an equivalent of the latter, and only ever uses the former — a nuance that is lost when official documents are translated in English. In other cases, linguistic confusion is brought in by omissions of words and concepts that are given for granted in some cases and not in others. Some definitions, for example, don't bother to specify the (obvious) fact that misconduct includes intentional acts, while others do. Therefore, one is left with the doubt that other concepts might also be omitted but given for granted. How many definitions, for example, implicitly include reckless acts? How many include the misuse of statistics, misrepresentation of credentials, or ghost authorship? Most problematic of all, however, is the case of the same words being used to mean different things. The disagreement between the NSF and NAS over the "other serious deviations" clause, for example, has been explained as a misunderstanding, in which one party interpreted the sentence in purely ethical terms (i.e. deviation from acceptable *ethical* practices) and the other interpreting it exclusively with research practice [28].

3 TOWARDS A UNIFIED DEFINITION

A simple and quick way to homogenise definitions across international borders could be to converge on a "minimum common denominator" of behaviours recognised as misconduct in all countries. As Tables 1 and 2 suggest, this would be easily achieved by narrowing the definition to FFP (or an equivalent category), with the culpability bar set at intentional acts. This approach might be appropriate when investigating allegations in international collaborations, especially if countries with narrow definitions like the USA are involved [5]. However, as explained above, such a restricted definition would not meet the other objectives that a unified definition could and should have. It would do little to protect the scientific literature from bias, would provide no guidance to foster good research throughout the world, and would not promote higher values in science.

A more comprehensive approach might consist in developing a broad but articulated definition, in which conceptual distinctions are made among types of behaviours. Resnik (2003), for example, proposed a new, more complete definition of misconduct as the "intentional violation of accepted scientific practices, common-sense ethical norms, or research regulations" and then specified that only FFP and interfering with misconduct investigations were "punishable" [2]. Resnik's definition does not solve, in my opinion, most of the ambiguities and limitations discussed in this paper, (for example, in its use of terms like "serious", "accepted", "common-sense", "questionable", etc.). These ambiguities might be removed by completely alternative approaches to defining misconduct — for example that proposed by the CRI in 1995, which for unclear reasons went completely ignored. Such alternatives should be explored, developed and considered further.

In conclusion, a consensus on "core" research misconduct is ready at hand, but the global fostering of research integrity would benefit from a more comprehensive definition — one that encompasses not only the "black", but also the "grey" areas, and maybe the "white" ones too. To meet this objective, the international dialogue could fruitfully focus on the key concepts, aims and assumptions that underlie the definition itself.

References

[1] Steneck, N.H. (2006). Fostering integrity in research: Definitions, current knowledge, and future directions. *Science and Engineering Ethics*, 12(1): 53–74.

[2] Resnik, D.B. (2003). From Baltimore to Bell Labs: Reflections on two decades of debate about scientific misconduct. *Account Res*, 10(2): 123–135.

[3] Resnik, D.B. (2009). *International Standards for Research Integrity: An Idea Whose Time has Come? Accountability in Research: Policies and Quality Assurance*, 16(4): 218–228.

[4] OECD–GSF (2007). *Best Practices for Ensuring Scientific Integrity and Preventing Misconduct*, OECD — Global Science Forum.

[5] OECD–GSF (2009). *Investigating Research Misconduct Allegations in International Collaborative Research Projects — A Practical Guide*, OECD — Global Science Forum.

[6] ESF (2000). Good scientific practice in research and scholarship. E.S. Foundation, Editor.

[7] ESF–ORI (2007). *Research Integrity: Global Responsibility to Foster Common Standards*, European Science Foundation — Office of Research Integrity (US).

[8] Boesz, C. and Lloyd, N. (2008). Collaborations: Investigating international misconduct. *Nature*, 452(7188): 686–687.

[9] Ankier, S. (2002). Dishonesty, misconduct and fraud in clinical research: An international problem. *Journal of International Medical Research*, 30(4): p. 357–365.

[10] Fanelli, D. (2010). "Positive" results increase down the hierarchy of the sciences. *PLoS One*. 5(3).

[11] Zammito, J.H. (2004). *A Nice Derangement of Epistemes: Post-Positivism in the Study of Science from Quine to Latour*. Chicago: Chicago University Press.

[12] Bosch, X. (2010). Safeguarding good scientific practice in Europe. *EMBO Reports 2010*, 11(4): 252–257.

[13] Guston, D.H. (1999). Changing explanatory frameworks in the US government's attempt to define research misconduct. *Science and Engineering Ethics*, 5(2): 137–154.

[14] Schachman, H.K. (1993). What Is Misconduct in Science. *Science*, 261(5118): 148–149.

[15] NAS (1992). Responsible science, Volume I: Ensuring the integrity of the research process. *Committee on Science, Engineering, and Public Policy, Panel on Scientific Responsibility and the Conduct of Research.* Washington, D.C: National Academy of Science.

[16] DHHS (1995). *Integrity and Misconduct in Research — Report of the Comission on Research Integrity*. Department of Health and Human Services — Public Health Service.

[17] PHS (2005). *Public Health Service Policies on Research Misconduct*. Department of Health and Human Services — Public Health Service, Federal Register, 70(94).

[18] Spece, R.G. and Bernstein, C. (2007). Scientific misconduct and liability for the acts of others. *Medicine and Law*, 26: 477–491.

[19] Nylenna, M., *et al.* (1999). Handling of scientific dishonesty in the Nordic countries. *Lancet*, 354(9172): 57–61.

[20] NCISM (2010). *The National Commission for the Investigation of Scientific Misconduct*. Accessed on 26 March 2010 [cited 20 August 2010]; Available from: http://www.etikkom.no/en/In-English/Scientific-Misconduct/.

[21] TENK (2002). *Good Scientific Practice and Procedures for Handling Misconduct and Fraud in Science*. Finnish National Advisory Board on Research Ethics.

[22] DCSD (2006). *Annual Report 2005*. Danish Committees on Scientific Dishonesty — Danish Agency for Science, Technology and Innovation.

[23] Anonymous (2003). More heat, less light on Lomborg. *Nature*, 421(6920): 195.

[24] Abbott, A. (2003). Social scientists call for abolition of dishonesty committee. *Nature*, 421(6924): 681.

[25] Andersen, H. (2007). *Demarcating Misconduct from Misinterpretations and Mistakes.* First Biannual SPSP Conference. Twente.

[26] DRA (2003). *Report on the Rules Governing Research Ethics.* Danish Research Agency.

[27] DASTI (2009). *Executive Order on the Danish Committees on Scientific Dishonesty — Consolidated Act* (Unauthorised translation), Danish Agency for Science, Technology and Innovation — Ministry of Science, Technology and Innovation.

[28] Buzzelli, D.E. (1993). The Definition of Misconduct in Science — a view from NSF. *Science*, 259(5095): 584–585.

[29] NHMR–ARC (2007). *Australian Code for the Responsible Conduct of Research.* A.R.C. National Health and Medical Research Council (eds.).

[30] CAS. [cited 06 July 2010]; Available from: http://www.las.cas.cn/jypx/yjsjy/xzfc/yjsdzb/200909/P020090921309232652282.doc.

[31] CESHE (2007). *Ethics Code.* Croatian Committee for Ethics in Science and Higher Education.

[32] Breittmayer, J.P., *et al.* (2000). Responding to allegations of scientific misconduct: The procedure at the French National Medical and Health-Research Institute. *Science and Engineering Ethics*, 6(1): 41–48.

[33] ICMR (2006). *Ethical Guidelines for Biomedical Research on Human Participants.* Indian Council of Medical Research.

[34] ALLEA, *et al.* (2003). *Memorandum on Scientific Integrity.* ALLEA — ALL European Academies.

[35] MER (2006). Act of 30 June 2006 No. 56 on ethics and integrity in research, Ministry of Education and Research — Norwegian Government.

[36] SRC (2004). *The Swedish Research Council's Expert Group for Investigation of Suspected Research Misconduct — Guidelines for the Group's Work.* Swedish Research Council.

[37] SAAS (2008). *Integrity in Scientific Research — Principles and Procedures.* Swiss Academies of Art and Sciences.

[38] UKRIO (2009). *Code of Practice for Research: Promoting Good Practice and Preventing Misconduct.* London: UK Research Integrity Office.

CHAPTER 15

KEYNOTE ADDRESS: PROMOTING INTEGRITY IN RESEARCH

David L Vaux

The term "research integrity" is used in two ways, to refer to the integrity of the scientific literature (i.e. its freedom from errors), and to refer to the integrity of scientists (i.e. the consistency with which they adhere to agreed principles and codes).

1 INTEGRITY OF THE SCIENTIFIC LITERATURE

For science to be useful, and for it to progress efficiently, it is necessary for the errors that will inevitably enter the scientific literature to be detected, reported and corrected rapidly. Errors can enter the scientific literature in three ways:

Firstly, all scientific data is subject to random errors, so anomalous results will always occur. With the use of computers and large data sets, combined with publication bias towards positive results, and over-reliance on statistical significance reported as P values, this problem has been growing, especially in the life sciences. This has led some to conclude that more than 50% of published research findings are false (Ioannidis, PLoS Med. 2005; 2: e124), and explains, for example, why many claimed genotype-disease associations have not been replicated (e.g. Morgan *et al.*, JAMA. 2007; 297: 1551–1561).

Secondly, errors can result from carelessness or incompetence. If care is not taken, arithmetical mistakes can be made, or the wrong statistical tests used, or a wrong image file might be incorporated. Authors can forget to label the axes on graphs, or describe the error bars in figures (e.g. Vaux, Nature. 2004; 428: 799). Failure to describe error bars, applying statistics

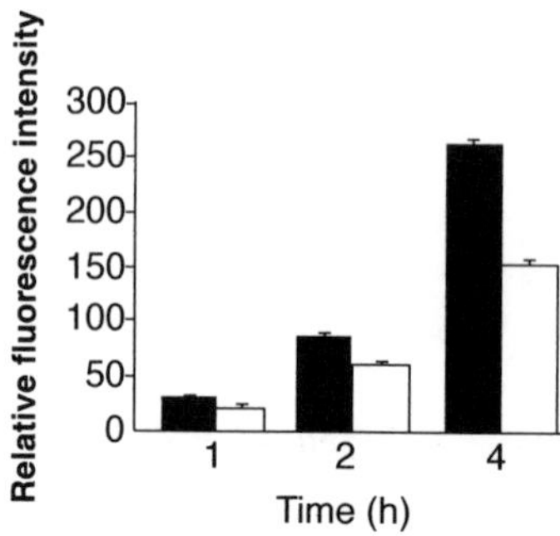

Fig. 1. Errors due to incompetence, bias, & sloppiness. Statistics should not be shown when N = 1.

to replicate results rather than using independent samples, and making arithmetical errors, might occur in as many as 30% of papers in the biomedical sciences (García-Berthou and Alcaraz, BMC Med Res Methodol. 2004; 4: 13).

Thirdly, errors can arise as a consequence of research misconduct, that is, due to deliberate falsification, fabrication, or misrepresentation of data. Anonymous surveys have suggested this may occur in around 1% of publications and often goes unreported (Titus *et al.*, Nature. 2008; 453: 980–982).

No matter how errors arise, or what measures are taken to reduce their number, for science to advance efficiently, it is important that these errors are found, reported, and the publications corrected. This requires readers of papers who spot errors, and researchers who cannot replicate results, to take action, either by contacting the authors of the paper, the host institution, the journal's editors, or an integrity office or ombudsman. It also requires action on the part of those who receive the report to verify that the error exists, and have a correction published.

It is important to note that only the third source of errors is a consequence of misconduct, because only these are deliberate. No one would deliberately try to look careless, make spelling mistakes, or not label their figures.

Even though scientific misconduct requires intent, it is not all of equal severity, but ranges from the mild (sometimes referred to as "questionable scientific practices", such as deliberately failing to cite competitors' works), to the moderate (such as cropping the image of a gel to remove cross-reactive bands or empty lanes), to the serious (e.g. fabricating data).

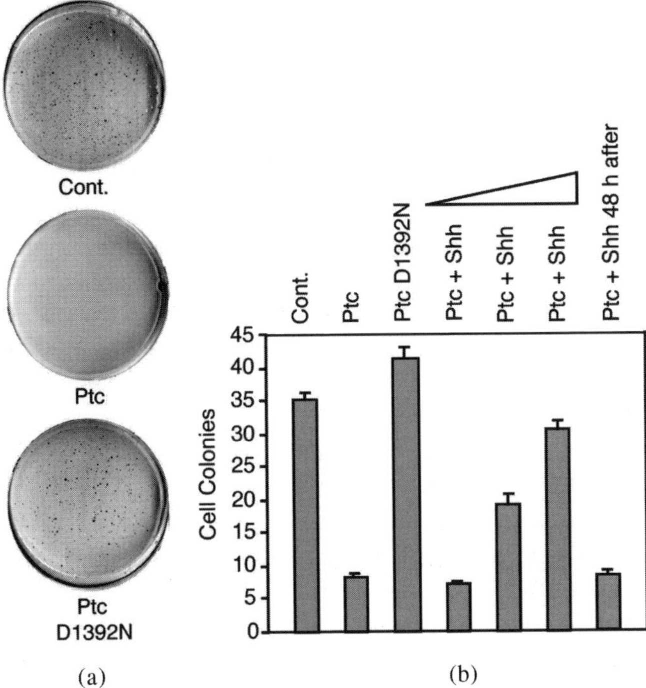

Fig. 2. Ptc expression inhibits growth in soft-agar assays through apoptosis induction. 293T cells were transiently transfected in the presence or absence of Ptc-1 (wild-type or mutant) with or without Shh. Shh was either added simultaneously to the plating in soft agar or added 48 hours later. Cells were allowed to grow for 12 days in soft agar. (A) A set of representative plates is shown. (B) The colony number was determined as the number of isolated clones with a diameter larger than 100 μm. Errors bars indicate SDs (n = 3).

Research misconduct is a problem not only because it can cause errors to enter the scientific literature, it is a behavioural issue that can undermine confidence in the scientific enterprise. Therefore, not only would errors in the literature due to misconduct need to be corrected, procedures need to be put in place to impose appropriate sanctions on the offending researchers.

2 INTEGRITY OF RESEARCHERS

Fostering the integrity of researchers is not only important to reduce the number of errors entering the literature, it will also help minimise harm to humans, animals, and the environment by promoting adherence to laws, regulations, and ethical codes, as well as encourage competition to be fair, with credit given where it is due.

Appropriate and fair allocation of credit is particularly important for scientists, because it not only determines their standing amongst their peers, but also their ability to attract funding for their salaries and to carry out their research.

Plagiarism (the use of someone else's words or ideas without the proper citation or acknowledgement) is an example of misallocation of credit. It is generally included under the definition of "scientific misconduct", but unlike data fabrication and falsification, plagiarism does not usually result in errors of scientific facts entering the literature. However, as it involves false attribution of credit, it can erode the puplic's confidence in research as a career, especially if it is tolerated.

In addition to plagiarism, there are other practices that involve false allocation of credit. For example, there is honorary authorship, in which an author is listed on a publication, even though they have made no substantial intellectual contribution to it; there is ghost authorship, in which an author of a paper is not listed, and there are excluded authors, in which researchers who have made a contribution that would justify inclusion as an author are nevertheless not listed.

The incidence of inappropriate authorship is difficult to determine, but in clinical journals, it has been reported to occur in as many as 19% of publications (Flanagin *et al.*, JAMA. 1998; 280: 222–224).

3 PROMOTING RESEARCH INTEGRITY

Promoting the integrity of researchers requires both a bottom-up "values" based approach, as well as top-down mechanisms to ensure compliance.

Of course, scientists have always discussed and applied their own values and ethics, but having a written set of agreed principles, guidelines and codes of conduct can help establish consensus for what is acceptable. Regulations and licences are useful for ensuring appropriate human and animal ethics and occupational safety practices.

But the mere existence of codes *per se* is unlikely to alter behaviour. Mechanisms to ensure compliance are also necessary.

The development of such codes and ensuring compliance is not only the responsibility of researchers themselves, but also depends on other bodies, including the host institution where the research is performed, funding bodies, journals, learned bodies and scientific academies, and ombudsmen and offices of scientific integrity.

There are two general ways of promoting compliance: the inspection/police force model, and the whistle-blower/fire alarm model.

The fire alarm model, in which researchers report concerns of possible errors or possible misconduct, allows those with the most knowledge of the norms of the relevant discipline to raise a concern and have it addressed. This model can be applied to all published research, including those funded privately or commercially. However, this model requires the cooperation of institutions to investigate the allegations, the funding bodies to withhold funding, and the journals to publish corrections or retractions. Complex systems tend to operate best when there is independent oversight provided at each level. Ombudsmen and offices for research integrity can provide this oversight, as well as help revise codes, collect data, provide advice, and act as an avenue for appeal.

Maintaining the integrity of the scientific literature, and fostering the integrity of researchers, will be helped if all bodies accept their responsibilities.

Individual researchers should make themselves aware of the relevant codes, guidelines and regulations. If they come across possible errors or misconduct, they should report it.

Institutions should have codes of conduct, and should put in place protocols for handling reports of possible errors or misconduct.

Journals should implement screening of accepted manuscripts to lessen the incidence of errors persisting through to publication. They should have guidelines to help authors and reviewers. They should have protocols for handing reports of possible misconduct or errors. They should readily publish corrections, rebuttals, and failure to replicate earlier studies, and do so in a way that the correct paper is cited, rather than the original. They should publish policies on authorship and declarations of conflict of interest.

Funding bodies should cooperate with ombudsmen and offices for research integrity so that financial sanctions can be applied when misconduct is proven.

Learned bodies and scientific academies should provide leadership by example, and help craft codes of conduct that are appropriate for different disciplines.

Learned bodies, scientific academies, funding bodies, institutions and individual researchers should push for the establishment of ombudsmen or offices for research integrity where none exist.

CHAPTER 16

DOES PEER REVIEW WORK AS A SELF-POLICING MECHANISM IN PREVENTING MISCONDUCT: A CASE STUDY OF A SERIAL PLAGIARIST

Ben R Martin

1 INTRODUCTION

Many fondly assume that "The Republic of Science" operates successfully on the basis of "self-policing". One of the implicit assumptions here is that research misconduct is rare, generally low-level and self-correcting. A second is that any serious misconduct is quickly detected by peer review and stopped. A third is that the risk of being caught and the severe repercussions that follow are such that few researchers are tempted to stray. However, all this presupposes that peer review does indeed succeed in detecting misconduct, and that editors, publishers, universities and the wider research community then work effectively together to investigate problem cases and implement any necessary sanctions. In what follows, I describe a case study demonstrating what happens when those involved *do not* work closely together, a case study that may force us to reconsider our cherished preconceptions about the efficacy of self-policing.

2 "THE TIP OF AN ICEBERG"

I serve as Editor of *Research Policy* (*RP*), a leading journal in the field of science policy and innovation studies. In 2007, an alert PhD student reported that an *RP* paper published 14 years earlier (Gottinger, 1993) bore a certain similarity to a 1980 article in the *Journal of Business* (Bass, 1980). The *RP* Editors therefore set out to investigate.

Our first hypothesis was that this might turn out to be borderline plagiarism (H1) — a few phrases that were similar, or maybe a failure to cite a reference acknowledging credit for some part of the methodology. However, comparison of the two papers revealed that the main sections were 80–90% identical (as was subsequently confirmed independently by two referees[1]).

Our second hypothesis was that *the author may have produced just a single paper, then disappeared from the world of academic publishing* (H2). However, a check on the Web of Science revealed that the author, Hans-Werner Gottinger,[2] had published nearly 120 items in the international journals covered by the database, an extremely impressive total for a social scientist.

This led next to a third hypothesis — *perhaps the RP paper represented a one-off "moment of madness"?* (H3) Given that the name "Gottinger" is a somewhat unusual one, we tried "Googling" "Gottinger + Plagiarism" on the off-chance. Somewhat to our surprise, this revealed that in 1999 the Editors of *Kyklos* (Frey *et al.*, 1999) had retracted a 1996 paper by Gottinger because it plagiarised a prior article by Wyatt (1992).

The RP paper had been published in 1993 and the Kyklos paper in 1996, so this led to a fourth hypothesis: that the problem of research misconduct occurred during a relatively brief period in the early to mid-1990s when the author may have been under stress (H4). If so, once he had been exposed in 1999, then presumably the misconduct would have immediately ceased. To investigate this, we began checking some of the papers he had published after 1999. On about the third one checked, a 2002 article in the International Journal of Global Energy Issues (Gottinger, 2002a), we found evidence suggesting that this heavily plagiarised a 1997 article (Chen, 1997) and duly informed the editor.

By this stage, the case was clearly getting serious. It involved a serial plagiarist, one who, more worryingly, had not stopped when caught by the editors of *Kyklos* in 1999. However, we had done our duty as journal editors

[1]In addition, one referee spotted that, besides plagiarising Bass (1980), Gottinger had also fabricated all the data used in his paper. As the referee wistfully remarked, "In for a penny, in for a pound!"

[2]There is only one academic author with the name "HW Gottinger" recorded in the Web of Science. This made the task of investigation much easier than if he had a common name like "J Smith".

in exposing the research misconduct, and it was now time to hand the case over to his employer for a full investigation of his other 100 articles and dozen books.[3]

3 THE SEARCH FOR HIS EMPLOYER

From the institutional address he had given in numerous publications over a 20-year period (e.g. Gottinger, 1987, 1996 and 2007) and from an online resumé,[4] it was apparent that Hans-Werner Gottinger was a professor in the Institute of Management Science (IMS) at Maastricht University in the Netherlands, where he had held a chair since 1983, serving as IMS Director from 1987–1990 and as Dean of Faculty from 1983–1987. We therefore searched the Maastricht University website but were puzzled to find that Gottinger was not listed among their faculty, nor was there any mention of the Institute of Management Science. But perhaps he had recently left the University? And perhaps the Institute had changed its name or been merged with another department? We therefore contacted the University to check. To our amazement, they reported back that Gottinger had never been an employee and that the University had never had an "Institute of Management Science". Alerted to this serious misrepresentation, they sent an official letter to Gottinger,[5] instructing him to remove this false institutional information from the websites and publications[6] in which it appears.[7]

[3]For a list of these books, see http://www.amazon.com/Hans-Werner-Gottinger/e/B001HPKOV2. Accessed on 4 October 2010.

[4]This was previously available at http://www.worldsustainable.org/conferences/conf3_keynote.html, accessed in June 2007; it has since been removed from this webpage, but an archived version can be found on www.archive.org. Note that his claim in this to be Associate Editor of five named journals seems to be false in several if not all instances.

[5]See http://www.volkskrant.nl/archief_gratis/article592340.ece/We_kennen_die_Hans_Gottinger_echt_niet. Accessed on 4 October 2010.

[6]Unfortunately, this is not so simple. Once an article has been published giving a particular institutional address, it cannot be simply "recalled" to be corrected. And while websites can be corrected, earlier versions often continue to exist in a cached or archived form. Hence, three years later, the false institutional affiliation can still be found in numerous places.

[7]In his response to the Maastricht letter, Gottinger's explanation was that "I was associated for several years with EIPA [European Institute of Public

In several other articles, Gottinger (e.g. 1996 and 2004) stated that he was employed at the impressive-sounding International Institute for Environmental Economics and Management (IIEEM), at Bad Waldsee in Germany. However, a check on Google found no website for such an institute; indeed, according to Google, the only person ever to have listed this address was Gottinger. In more recent papers (e.g. Gottinger, 2000), Gottinger's institutional address was given as the International Institute of Technology Management and Economics (IITME), where, according to his CV, he was Director for four years. Another check on Google found no website for this institute, and again the only researcher ever listing this as his institutional address was Gottinger. Indeed, the postal address later given for this institute[8] turned out to be identical with Gottinger's home address as recorded in the German telephone directory.[9,10]

Administration], Maastricht and MSM [Maastricht School of Management, a quite separate institution from the University], and this was inadvertently extended in the context of a few publications... [I] sincerely apologise for this mishap." Apparently, for 20 years he *thought* he was a Professor in the Institute of Management Science in Maastricht University, including a period of three years when he thought he was Director of that Institute, and another period of four years when he thought he was Dean of the Faculty of Economics in Maastricht University. But it turns out he was actually associated with two entirely different institutions all along — a simple "mishap"! Except, of course, it does not seem he was ever employed by MSM, and EIPA say they have no record of him either.

[8]For example, in 2007, Gottinger was listed among the Advisory Editors of the *International Journal of Revenue Management*, where IITME's address was given as Unterring 21, 85051 Ingolstadt, (www.inderscience.com/browse/index.php?journalCODE=ijrm — downloaded in July 2007; since removed but still accessible via www.archive.org).

[9]See http://www.11880.com/telefonbuch/index/search?method=searchSimple&location_id=&_dvform_posted=1&name=Gottinger&firstname=Hans-Werner%20&zipCity=PLZ%20oder%20Ort&street=Stra%C3%9Fe&streetNumber=Nr. Accessed on 4 October 2010.

[10]Among the numerous other suspect institutional addresses used by Gottinger over the years on papers and at conferences are: Institut für Mathematische Writsaftsplanung [sic], Schlos [sic] Rheda, Rheda, BRD (see Gottinger, 1976a & b, two identical papers — an apparent case of self-plagiarism); IIRRM, Schloss Waldsee, Bad Waldsee (presumably yet another "International Institute" — given as his address when Gottinger was still listed as an Associate Editor of the *International Journal of Environment and Pollution* in 2007); the International Institute for Information Technology, Maastricht (see http://www.iamot.org/IAMOT2001/PDF/Papersess. pdf, accessed on 4 October 2010); the Department

4 SO HOW WELL DOES SELF-POLICING WORK?

Given that there was no obvious university employer to pursue the investigation, we continued to explore on the web and through other channels what could be found out about the individual at the heart of this case. Fortunately for us (and unfortunately for him), the fact that he had an unusual name[11] made this comparatively straightforward. This extended investigation produced disturbing evidence on how ineffectively self-policing works in other situations within the world of research and academia.

While one might not be entirely surprised to learn that peer review occasionally slips up with regard to ensuring the research integrity of individual publications, our fifth hypothesis was that *peer review surely works much more thoroughly when it comes to the appointment of individuals to senior academic positions* (H5). For example, when a university department is looking to appoint a new Chair of that department, presumably they will leave no stone unturned in checking to ensure that not only do they have the best candidate but also that the chosen person is *bona fide*.

In 1999, Rensselaer Polytechnic Institute (RPI) were searching for a top economist to head their Economics Department. Having gone through the conventional selection process, they appointed Gottinger. Shortly afterwards, a diligent PhD student at RPI (PhD students are the unsung heroes of this particular investigation!) came across the recent retraction by *Kyklos* of Gottinger's 1996 paper and informed the university authorities. Gottinger was summoned to account for this, but his "explanation" was rejected and he was forced to resign almost immediately, although he continued to use his RPI email address for many years after.[12]

Surely *peer review works even more thoroughly when a university comes to choose its new Rector, Vice-Chancellor or President* (H6). This was our sixth hypothesis, arguably constituting the critical test of the ability of

---

of Meteorology, Istanbul Technical University (e.g. Gottinger, 2002b, along with another nine articles all by Gottinger in the same "bumper" issue of the journal); and the Institute of Economic Analysis, University of Osaka, Osaka (e.g. Gottinger, 2005).

[11]The search so far has revealed no one else with the name "Hans-Werner Gottinger". There is, however, another person named "Hans Göttinger" (the "o" being spelt in this case with an umlaut), a medical doctor in Germany.

[12]See e.g. http://globalcommunitywebnet.com/gdufour//2004WorkGottinger.htm. Accessed on 4 October 2010.

peer review to detect misconduct. In 1993, the University of Klagenfurt in Austria was seeking to elect a new Rector. From all the candidates, and after obtaining references and taking expert advice, they narrowed this down to a shortlist of just three. Included on this list was Prof. Dr. Hans Gottinger, who, they had seemingly established, was from the University of Maastricht.[13] In the first round of voting, he came second, while in the run-off, he more than doubled his votes, coming within six votes of winning. Throughout the process, Klagenfurt University was apparently unaware that his Maastricht affiliation was bogus.

Clearly, no other university would make that mistake — or would they? In 1998, the University of Salzburg was looking to appoint a new Rector. After reviewing the candidates and making all the necessary checks, they drew up a shortlist of eight candidates. Among them was Prof. Dr. Hans Gottinger, who (they had established to their satisfaction) was "Professor of Managerial and Environmental Economics, University of Maastricht".[14] Or so they thought.

5 THE CASE GOES PUBLIC

Once the seriousness and complexity of the case became apparent, we invited the journal *Nature* to carry out their own investigations, specifically to ensure that all the facts were independently double-checked. These investigations were led by *Nature*'s German editor, Alison Abbott, who with her colleagues was able to pursue enquiries not only in Europe but also in the United States and in Japan (where Gottinger had been employed in recent years). By August 2007, the essential facts of the case had been firmly established, so *Nature* published an exposé (Abbott, 2007) while *RP* published an editorial (Martin *et al.*, 2007).

What happened next? Suddenly, from all round the world we received a flood of new information about Gottinger. The number of confirmed cases of plagiarism rose from three to 14. Of these, nine were undetected by referees and resulted in publication before the plagiarism was subsequently detected; only five were caught prior to publication. In other words, peer

[13]See http://www.uni-klu.ac.at/home/mitteiblatt/old/94-95/mittei31. Accessed on 4 October 2010.

[14]Press release issued by University of Salzburg, 19 May 1998, entitled "Acht Kandidaten für Rektorswahl".

review would not appear to have a very successful "hit rate" when it comes to detecting plagiarism *before* publication.

In addition, we received evidence of around half a dozen instances where Gottinger had been fired or forced to resign. Only in some of these was this due to reasons of plagiarism. On two separate occasions, he seems to have held two full-time jobs and drawn two salaries at the same time, without either of the employers involved being aware of this. First, during the period 1976–1979 he held a full-time chair at Bielefeld University (where he had been appointed back in 1973) and another chair at Groningen University.[15] When the latter found out, he was forced to leave (although Bielefeld University were unfortunately not informed). Gottinger then went and secured another second job, this time at the National Research Centre for Environment and Health (GSF) in Southern Germany. Bielefeld University only found out about this a year later in 1980, when he was forced to quit (Abbott, 2007). According to investigations by *Nature*, Gottinger was subsequently dismissed by GSF in 1984 for submitting a research proposal to the European Commission in which he had allegedly forged letters of support from a number of companies (*ibid.*). Later, Gottinger served as Director of the prestigious Fraunhofer Institute for Technological Trend Analysis from 1988 to 1990. However, when a book he had published a few years earlier in 1983 was discovered to have plagiarised a 1974 US government report, he was asked to leave (Anon, 2007).

The deluge of further information about Gottinger and his career enabled us to address our seventh and final hypothesis about peer-review and its efficacy in "self-policing" the academic community. This is that, *while an individual might be able to get away with plagiarism in a few cases over short period of time, the self-policing mechanisms of the academic community will ensure that he or she cannot continue with this over a prolonged period, let alone make a career out of it* (H7). From the information we received after publicising the case in August 2007, it became apparent that, rather than becoming involved in plagiarism late in his career in the 1990s, as we had first assumed, Gottinger had been engaged in acts of plagiarism for over 30 years — virtually his entire academic career. The first confirmed case took place in 1974. It came to light in 1978, when Gottinger was forced to admit in the journal *Automatica* that he had taken

[15]See http://www.rug.nl/bibliotheek/collecties/ub/gamma/bedrijfskunde?lang=en. Accessed on 4 October 2010.

"essentially all ideas, methods and conclusions" from a 1974 paper by Ho and Chu (see Gottinger, 1978). Indeed, Gottinger evidently liked the Ho and Chu paper so much he plagiarised it not once but three times in separate journals (*Annals of Systems Research*, 1974; *Economic Computation and Economic Cybernetics Studies and Research*, 1975; and *Revue Française d'Automatique, Informatique, Recherche Opérationnelle* (*RAIRO*), 1976 — see Gottinger, 1978).[16] This is incredibly reckless behaviour, even for a plagiarist. In plagiarising the same article three times, was he half hoping to get caught and stopped? Was he perhaps testing the system? What is important to note is that all that he received for this act of triple plagiarism was a very gentle "slap on the wrist" — the low-key retraction published in *Automatica* was labelled an "acknowledgement of priority" rather than an "admission of plagiarism". Not surprisingly, therefore, it had little impact at the time, either on Gottinger or on the wider research community. Fortunately, one person remembered it and drew it to our attention 30 years later.

6 QUESTIONS RAISED BY THE CASE

This misconduct case raises a number of troubling questions. One is, "How many of the other 100 published journal articles and dozen or so books published by Gottinger are based on plagiarism?" As noted above, we have so far obtained evidence of nine cases where an article or book was published and plagiarism was only detected subsequently, compared with just five cases where the plagiarism was caught *before* publication by alert referees or editors. As for all his other numerous publications, we just do not know how many are "genuine" and how many involve plagiarism. Until each has been checked, they must all be treated as potentially suspect. The difficulty here is that checking all of them would require an immense effort, not least because in some cases "the author" was cleverer in trying to disguise the source plagiarised. Instead of copying journal articles, he took an apparent fancy to doctoral dissertations, presumably on the assumption that these were unlikely to be published so the probability of being caught was slight. In particular, he drew upon PhD theses from leading US universities

[16] Despite this admission that Gottinger (1974) was based on plagiarism of Ho and Chu (1974), this did not deter Gottinger from continuing to cite it (see Gottinger, 1980).

including Berkeley, Cornell and Harvard as sources for "his" papers (e.g. Gottinger, 2001).

A second question often asked is, "If he was making up all these bogus institutional affiliations, where did he actually get his income from?" As far as one can tell from his CV and other information collected, he was never out of a job for long, holding a succession of around two dozen posts over a career spanning nearly 40 years. Indeed, as noted earlier, on two occasions he held two jobs simultaneously, drawing two salaries. So income was not apparently a problem. Whenever he was sacked or resigned from one position, he seemingly encountered little problem in finding another, although over time he was forced to go progressively "down-market" in terms of the status of the institutions for which he worked.[17]

Third, "How did he manage to evade detection for so long?" There are various possible explanations. First, several people have described him as something of "a loner", and for nearly all his publications he was the sole author, so there were no fellow authors checking his work and asking awkward questions. Moreover, many of his papers were highly mathematical (at least, for a social scientist), so referees and other readers may have tended to skip over the equations, thus failing to spot that they were based on plagiarism. However, a major part of the explanation of how he managed to evade getting caught may lie in the bogus institutional affiliations he gave. According to those papers where Gottinger listed it as his institutional address (e.g. Gottinger, 1987), the Maastricht "Institute of Management Science" had a different PO Box from Maastricht University (the former was given as PO Box 591, while the latter has as its address PO Box 616, Maastricht). Any request to the "IMS Director" for a reference when Gottinger had applied for some other job could therefore presumably be intercepted at the post office and a suitable "reference" provided. Likewise, any request for a reference addressed to the International Institute for Environmental Economics and Management (IIEEM), Schloss Waldsee, 88339 Bad Waldsee, Germany, would end up in the pigeon-holes at the Gesellschaft für Internationale Studien mbH, the international school located at that address in the 1990s, when Gottinger was temporarily employed there as a lecturer. Similarly, correspondence

[17]For a large part of the first decade of the 21st Century, he worked at the Institute of Management Science at Kwansei Gakuin University, a lesser-known Japanese university.

addressed to the International Institute of Technology Management and Economics, Unterring 21, 85051 Ingolstadt, would be duly delivered to his private home. All of this suggests careful planning as well as premeditation.

Furthermore, Gottinger listed a number of extremely impressive referees on his CV, including one Nobel Prize laureate and another who was president of a national Academy of Science. Many had known Gottinger for 20–30 years and, when informed of his serial plagiarism in the autumn of 2007, they expressed astonishment as well as great sadness.

A fourth essential question is this: "When he was caught engaging in research misconduct in the past, why was he not stopped?" It is widely assumed that self-policing not only ensures that offenders are "exposed" but also that they are shamed into stopping or, in more serious cases, are denied any further opportunity to continue their research and to publish. However, that is not what happened here. Whenever someone detected plagiarism or other misconduct, he or she assumed it was a "first offence" and they therefore let him off with a mild rebuke. As far as one can tell, the 1978 retraction by Gottinger in *Automatica* of his three earlier papers attracted virtually no attention, while the 1999 retraction by the editors of *Kyklos* attracted little more. The institutions that sacked him or forced him to resign were generally so embarrassed at having been hoodwinked that the dismissal was handled very quietly without the wider community being informed. So Gottinger continued in his merry ways.

Several individuals later admitted that they had harboured suspicions but unfortunately they assumed that "someone else" would clear up the problem. This can be seen as a form of "the tragedy of the commons" (Hardin, 1968). It may involve too much effort on the part of an individual to pursue their suspicions; however, in the long term, this only makes the situation worse for the academic community as a whole. Results from studies by economic game theorists based, for example, on "the dictator game" or "the ultimatum game" show that players tend to divide into two main groups — cooperative observers of the rules, and "free-riders" or "defectors". The latter group flourishes unless some of the former volunteer to take on the role of "punishers". Yet this often comes at a significant personal cost, so relatively few co-operators are willing to mete out punishment; most by default become "second-order free-riders" (e.g. Panchanathan and Boyd, 2004), leaving the job of punishing to someone else — which is precisely what occurred here.

Fifth, "Can plagiarism be considered as a 'victimless crime' and therefore one that is less serious than, say, data fabrication or falsification?" There are certainly some researchers who seem somewhat more relaxed about plagiarism. They assume that plagiarists will not risk publishing anything significant, with the result that they are seldom cited and therefore have little impact. Moreover, by lifting material from an obscure source, they may even assist in diffusing it and hence in giving it more impact than it would otherwise have had. However, there are strong counterarguments against this view. According to the Web of Science, Gottinger has earned approximately 300 citations,[18] many arguably "stolen" from other researchers. Those who learnt that their work had been plagiarised reported feeling "violated". Similarly, many of those unsuccessful candidates who came second to Gottinger when he obtained his various positions might have instead been appointed. In addition, referees, editors and publishers along with readers have all felt a sense of their trust being broken. Last but not least, vast amounts of effort are required to investigate complex misconduct cases such as this.

Sixth, "Is Gottinger a one-off — such a truly exceptional case that we do not need to matter too much about him?" The simple answer is that we do not know. However, we do know that the detection of serial plagiarism is on a sharp increase. There was just a single case reported in the 1980s, and another one in 1998; but in the last few years a dozen or more have come to light in science, social sciences and the humanities.[19] The same factors that account for the growing incidence of detected plagiarism among students (the growing availability of electronic material making it easier or more tempting to "lift" that material, and search engines making plagiarism far easier to detect) would seem to apply to academic researchers.

However, this leads on to a final question: "How many other serial plagiarists are 'out there' waiting to be discovered? How much plagiarism remains undetected?" Again, no one knows. There is, as we observed above, some information about the level of detected plagiarism, but, almost by definition, we have no idea about the level of *undetected* plagiarism.[20] This is

---

[18]Search on Web of Science carried out on 4 October 2010.

[19]See Martin *et al.* (2007, note 32), since when several other cases of serial plagiarists have come to light (e.g. Bouyssou *et al.*, 2009; Mashta, 2009).

[20]However, in the next section, we report some evidence that the level of undetected plagiarism may be many times greater than the level of detected plagiarism.

what the noted American philosopher, Donald Rumsfeld, would term "a known unknown" (quoted in Boardman, 2005, p. 783).

7 "THE CASE OF THE BITER BIT"

There is one remaining twist in the Gottinger tale. In nearly all of his more than 100 publications, Gottinger was the sole author, an unusual occurrence these days even among social scientists. However, in two articles, he had the same co-author. This led us to first wonder whether this co-author actually existed, or whether he was an "imaginary friend" created merely to give the illusion that Gottinger collaborated like everyone else. The co-author, it turns out, does exist. Was he aware of Gottinger's misconduct? To check involved "Googling" strings of words from the paper by Gottinger and Weimann (1992). This quickly produced a "hit" — large sections were virtually identical with those in a conference paper published by the Research Council of Zimbabwe (Salahuddin, 2004[21]). That seemed to confirm the plagiarism. Then came the sudden realisation that the latter had been published in 2004, 12 years *after* the 1992 paper by Gottinger and his colleague. In other words, the dates were the wrong way round. We had just discovered perhaps the world's first case where the work of an academic plagiarist had been plagiarised by a second plagiarist! (Just to be perfectly clear — there is absolutely no evidence that the 1992 paper is based on plagiarism; this may well be one Gottinger paper where no plagiarism was involved.)

Yet the ramifications of this accidental finding are decidedly troubling. What is the probability of a second plagiarist choosing one of Gottinger's papers as the source to plagiarise? If the overall occurrence of plagiarism in the research community is small, as most would assume, then the chances of this happening must surely be rather remote? One of the very few empirical studies on the extent of plagiarism among researchers[22] was conducted by Enders and Hoover (2006).[23] The findings from their survey

[21]The first half of this paper contain numerous strings of words identical to those appearing in Fox (1990), while the second half contains almost identical text to that in Gottinger and Weimann (1992).

[22]There are far more studies by academics on the prevalence of plagiarism among students than among academics.

[23]Another such study is that by Errami and Garner (2008), who found 73 "plagiarism candidates" in their database of 62,000 Medline abstracts; however, they

of 1,200 economists suggest that around 1 in 100 papers are plagiarised, although in the great majority of these the plagiarism is relatively minor (e.g. an unattributed phrase, idea or element of methodology). Perhaps only around a tenth of these — i.e. 1 in 1000 papers — involves whole-scale copying of the type exhibited here. However, we had examined only about 10 of Gottinger's papers before finding one plagiarised by someone else. The chances of this happening are only around 1% if 1 in 1000 papers are the subject of whole-scale copying. This should not have happened!

What are the possible explanations of this bizarre finding? First, it could just be a freak coincidence. Some might even choose to see it as divine retribution, or perhaps as proof that God does exist and indeed has a sense of humour! Second, it could imply that the actual occurrence of *previously undetected* plagiarism is approximately two orders of magnitude greater than the incidence of known or detected plagiarism. However, that would mean that only about 1 in 100 instances of plagiarism is detected through the normal peer review process, scarcely very reassuring. Thirdly, have we accidentally chanced upon an abnormal "hot spot" of plagiarism? It is, let us recall, a decidedly mathematical area where the referees may not check things properly. Yet the implication that peer review breaks down when confronted with more technical material is hardly very comforting, either. How may more such "hot-spots" are there? What proportion of science falls into this category where self-policing does not work?

8 CONCLUSIONS

Let us return to the question with which we started: does self-policing work with regard to academic misconduct? If self-policing of the research community operated in the way traditionally assumed, virtually none of this should have occurred. Gottinger should not have been able to get away with repeated plagiarism over a period of over 30 years. His use of bogus institutional affiliations should have been quickly detected and stopped. He should not have attained such a senior level in his profession, appointed as he was to the chair of the economics department in a leading university, and coming close to being chosen as Rector of not one but two universities. Indeed, the misconduct should have stopped when it was first detected

---

were only examining abstracts, so this merely provides a lower limit on the extent of plagiarism.

in 1978. Although he was caught on several subsequent occasions, each time those involved gave him the benefit of the doubt, assumed it was his first offence, and either let him off with a mild warning or quietly dispensed with his services without informing the wider research community. Hence in this particular instance, a life of academic "crime" does seem to have paid off.

What this case study reveals all too starkly is the need to revisit the assumption that self-policing works. If we are to reduce the future risk of such misconduct, greater vigilance and a willingness to pursue any well-founded suspicions are required on the part of *all* of us — whether referees, editors, publishers, employers or readers. Universities and other institutional employers need to investigate suspicious cases that have been brought to their attention, and to present the eventual findings to the wider academic community. We may also need to consider establishing an authoritative database where confirmed details of first-time offenders are registered so that others can check whether a particular suspect has been found guilty on a previous occasion and hence whether he/she is a "repeat offender" or not.

At a more personal level, if you have genuine suspicions about a publication or an individual, do not leave it to "someone else" to sort the problem out. If you do, you will be helping to create a new form of "the tragedy of the commons". While it may indeed involve a certain amount of effort to pursue a particular case, failure to do so will, in the long run, make the overall situation far worse for the academic community. The task of investigating and of "punishing" comes at a cost. Editors, universities and others must be prepared to incur that "cost" if the scourge of plagiarism and other research misconduct is to be kept at bay.

9 EPILOGUE

At least now that the various misdeeds have been fully exposed, that should finally have put paid to this particular case of misconduct. Not quite! In March 2008, a colleague "Googled" "Gottinger + 2008" and obtained a "hit". It was for the 5th Annual Future Business Technology Conference, organised by EUROSIS, the European Multidisciplinary Society for Modelling and Simulation Technology, where the Guest Lecturer, who was due to give an extensive 11-session tutorial course, was none other than Hans Werner Gottinger. According to the website, Gottinger is

"Professor of Managerial and Industrial Economics" and "Institute Director" of STRATEC at the Technical University of Munich.[24] Except, of course, he is not on the faculty of the Technical University of Munich, nor does the University actually have an institute called "STRATEC" (Abbott, 2008). Since then, he has continued to publish numerous academic books,[25] book chapters[26] and articles,[27] to speak at international conferences[28] and to serve on the editorial advisory board of journals,[29] not to mention running STRATEC Consulting[30] while claiming to be Research Director at IMD (the international business school in Switzerland) and at Sagentia (the international consultancy firm previously known as Generics)[31]. The impressive career continues.

---

[24]It was at http://www.eurosis.org/cms/index.php?q=node/354, accessed on 3 March 2008 and still available on 4 October 2010.

[25]These include one book on *Innovation, Technology and Hypercompetition: Review and Synthesis* (see http://www.routledge.com/books/details/97804154 94083/), another on *Strategic Alliances in Biotechnology and Pharmaceuticals: Sources and Strategies for Alliance Formation in a Global Industry* (see http:// www.startup-books.com/books/274984/Strategic-Alliances-in-Biotechnology-and-Pharmaceuticals%3A-Sources-and-Strategies-for-Alliance-Formation-in-a-Global-Industry/), a third on *Strategic Economics of Network Industries* (see https:// www.novapublishers.com/catalog/product_info.php?cPath=23_29&products_id=9746&osCsid=8dcecfb43aee90ba503663e09995dd7a), a fourth with the same main title as the second but no subtitle and now with two additional authors (see https://www.novapublishers.com/catalog/product_info.php?cPath=23_29&products_id=12459&osCsid=405f5a30a896632ef0a6a95caa3accf0), and a fifth on *Strategies of Economic Growth and Catch-Up: Industrial Policies and Management* (see https://www.novapublishers.com/catalog/product_info.php?products_id=20615&osCsid=), all accessed on 14 October 2010.

[26]See http://www.routledge.com/books/details/9780415469630/. Accessed on 4 October 2010.

[27]See Gottinger and Umali (2008b), which seems to be virtually identical to Gottinger and Umali (2008a) — an apparent case of self-plagiarism.

[28]See e.g. http://dipeco.economia.unimib.it/network/, and http://umconference.um.edu.my/globelics2010=3dc4876f3f08201c7c76cb71fa1da439, both accessed on 4 October 2010, and more recently http://www.icbmconference.net/icbm_accepted.pdf. Accessed on 29 December 2010.

[29]See http://www.bentham.org/open/tobj/EBM.htm, accessed on 4 October 2010.

[30]See http://stratec-con.com/. Accessed on 4 October 2010.

[31]See http://de.linkedin.com/pub/hans-gottinger/8/a58/811. Accessed on 4 October 2010.

10 ACKNOWLEDGEMENTS

This paper is based on a presentation made at 2nd World Conference on Research Integrity, Singapore, 21–24 July 2010. Earlier less detailed versions were published in Martin *et al.* (2007) and Martin (2007). The author is grateful to the Norwegian Centre for Advanced Study for support during the time when part of this investigation was being conducted. However, the main debt of gratitude is to other *RP* editors for their help in the investigation and to the dozens of individuals who provided so much information on this case, thereby helping to construct the overall picture reported here.

References

1. Abbott, A. (2007). Academic accused of living on borrowed lines. *Nature*, 448 (9 August), 632–633.
2. Abbott, A. (2008). The fraudster returns... *Nature*, 452 (10 April), 672.
3. Anon (2007). Mit fremden Federn geschmückt. *Frankfurter Allgemeine* (8 August 2007) (see http://www.faz.net/s/Rub7F4BEE0E0C39429A8565089709B70C44/Doc~EF8A9CC11F5DB4C66B36FB8425699228E~ATpl~Ecommon~Scontent.html, accessed on 4 October 2010).
4. Bass, F. M. (1980). The relationship between diffusion rates, experience curves, and demand elasticities for consumer durable technological innovation. *Journal of Business*, 53, S51–S67.
5. Boardman, M. E. (2005). Known unknowns: The illusion of terrorism insurance. *Georgetown Law Journal*, 93, 783–844.
6. Bouyssou, D., Martello, S., and Plastria, F. (2009). Plagiarism again: Sreenivas and Srinivas, with an update on Marcu. *4OR: A Quarterly Journal of Operations Research*, 7, 17–20.
7. Chen, Z. (1997). Negotiating an agreement on global warming: A theoretical analysis. *Journal of Environmental Economics and Management*, 32, 170–188.
8. Enders, W. and Hoover, G. A. (2006). Plagiarism in the economics profession: A survey. *Challenge*, 49, 92–107.
9. Errami, M. and Garner, H. (2008). A tale of two citations. *Nature*, 451, 397–399.
10. Fox, J. (1990). Symbolic decision procedures for knowledge-based systems. *Knowledge Engineering: Vol. II, Applications*. Adeli, H. (ed.), McGraw-Hill.
11. Frey, R. L., Frey, B. S., and Eichenberger, R. (1999). A case of plagiarism. *Kyklos*, 52, 311.

12. Gottinger, H. W. (1974). Information structures in dynamic team decision problems. *Annals of Systems Research*, 4, 1–20.
13. Gottinger, H. W. (1976a). Decomposition for stochastic dynamic systems. *Statistics*, 7, 889–903.
14. Gottinger, H. W. (1976b). Decomposition for stochastic dynamic systems. *Optimization*, 7, 889–903.
15. Gottinger, H. W. (1978). Acknowledgement of priority. *Automatica*, 14, 299.
16. Gottinger, H. W. (1980). Further results on decomposition for stochastic systems. *Optimization*, 11, 117–128.
17. Gottinger, H. W. (1987). Choice and complexity. *Mathematical Social Sciences*, 14, 1–17.
18. Gottinger, H. W. (1993). Estimating demand for SDI-related spin-off technologies. *Research Policy*, 22, 73–80 (now withdrawn because of plagiarism).
19. Gottinger, H. W. (1996). Competitive bidding for research. *Kyklos*, 49, 439–447 (now withdrawn because of plagiarism).
20. Gottinger, H. W. (2000). Book review: Competition in telecommunications. *Asia Pacific Journal of Management*, 17, 564.
21. Gottinger, H. W. (2001). Incentive compatible environmental regulation. *Journal of Environmental Management*, 63, 163–180 (now withdrawn because of plagiarism).
22. Gottinger, H. W. (2002a). Negotiation and optimality in an economic model of global climate change. *International Journal of Global Energy Issues*, 18, 181–201.
23. Gottinger, H. W. (2002b). Chapter 1: Issues of global environmental economics. *International Journal of Environment and Pollution*, 17, 424–434.
24. Gottinger, H. W. (2004). Econometric modelling, estimation and policy analysis of spill processes. *International Journal of Environment and Pollution*, 15, 333–363.
25. Gottinger, H. W. (2005). Economic growth, catching up, falling behind and getting ahead. *World Review of Entrepreneurship, Management and Sustainable Development*, 1, 101–120.
26. Gottinger, H. W. (2007). Competitive positioning through strategic alliance formation; review and synthesis. *International Journal of Revenue Management*, 1, 200–216.
27. Gottinger, H. W. and Umali, C. L. (2008a). Strategic alliances in global biotech pharma industries. *The Open Business Journal*, 1, 10–24.
28. Gottinger, H. W. and Umali, C. L. (2008b). The evolution of the pharma-ceutical-biotechnology industry. *Business History*, 50, 583–601.

29. Gottinger, H. W. and Weimann, P. (1992). Intelligent decision support systems. *Decision Support Systems*, 8, 317–332.
30. Hardin, G. (1968). The tragedy of the commons. *Science*, 162, 1243–1248.
31. Ho, Y. C. and Chu, K. C. (1974). Information structure in dynamic multi-person control problems. *Automatica*, 10, 341–351.
32. Martin, B. R. (2007). Research Misconduct — Does Self-Policing Work? In *Confluence: Interdisciplinary Communications* (Yearbook of the Centre for Advanced Study, Norwegian Academy of Science and Arts, Centre for Advanced Study, 2007/08) pp. 59–69.
33. Martin B. R. *et al.* (2007). Keeping plagiarism at bay — a salutary tale. *Research Policy*, 36, 905–911.
34. Mashta, O. (2009). Doctors try to get speakers to boycott international conference chaired by plagiarist. *British Medical Journal*, 339, b35–b45.
35. Panchanathan, K. and Boyd, R. (2004). Indirect reciprocity can stabilize cooperation without the second-order free rider problem. *Nature*, 432, 499–502.
36. Salahuddin, A. (2004). Intelligent decision support systems: Technology and applications. In *Impact of Innovative Science & Technology on National Wealth Creation*, Research Council of Zimbabwe. pp. 114–124. Proceedings of the 7th Science and Technology Symposium, Harare, 1–3 September (downloaded from http://www.symposium.rcz.ac.zw/7th_Symposium_Proceedings.pdf on 6 July 2007; since removed but still available on www.archive.org).

CHAPTER 17

SCIENTIFIC FALSIFICATIONS IN AND OUT OF SCIENCE

Edward P Kruglyakov

The problem of falsifications in scientific studies and consequences of these falsifications have become more and more serious, along with problems resulting from the growth and influence of pseudo-science. These problems affect even the most developed countries in the world. The main reasons for these increasing phenomena may be due to the scientific ignorance of senior bureaucrats, and of course, possibly due to corruption. The old story of telepathy studies in the United States and the Soviet Union to communicate with submarines is well known. In recent years, studies on anti-gravitation have also been supported. In this sense, the abnormal secrecy of defence-related research can create favourable conditions for pseudo-science to exist.

This paper gives some examples of the pseudo-science activity and falsifications in science starting from the last years of the Soviet Union and the first years of the Russian Federation until now.

Main features of the Russian Federation of the 1990s

The disintegration of the Soviet Union, the confusion experienced by the people, and uncertainty in the future created a very favourable environment for the growth of pseudo-science, collapse of morality, etc. The main processes of that time could be characterised as follows:

1. Corruption of high-ranking officials (sometimes together with scientific ignorance). As a result , such officials began to support astrology and such schemes as anti-gravitation, extraction of energy from stones, production of gold from cheap materials, the search for oil, gas and minerals with the use of non-existent physical fields (like torsion and micro leptons), application of senseless medical devices for the treatment of illnesses, etc.

2. Publicity to decieve people through the mass media coupled with the simultaneous lowering of the education level within the country;
3. Absence of moral principles of people in advertising, who often operated without legal limitations; and
4. Opportunity for pseudo-science "achievements" to be recognised in patents, in a variety of business activities, especially in medicine.

However, it should not be supposed that the pseudo -science is a particularly intrinsic feature of Russia. Traces of it may be found in the most developed countries.

1 FALSIFICATIONS IN SCIENTIFIC JOURNALS

These may be infrequent but they have grave consequences. Some of these are detailed below.

A. Okhatrin's paper about micro leptons (Doklady Akademyi Nauk, 1989): Some time later he made a photograph of a man in micro lepton radiation through 4 brick walls! Then a device was made that claimed to treat a great number of illnesses, as well as being able to prevent pathogenic radiation. Several years ago, a group of Okhatrin followers tried to obtain a license for oil exploration in the United Kingdom with the aid of micro lepton technology.

A. Akimov's review of torsion fields and their abilities (Biofizika, 1995): Torsion fraud is one of the greatest frauds that occurred in the last years (1989) of the Soviet Union and in the early Russian Federation. Initially, the torsion swindlers promised a great number of military applications such as:

- putting enemy troops into a disorganized, helpless state; and
- undetectable communication with satellites, submarines, secret service agents, etc.

At present, they promise:

- movement on the basis of an energy extraction from a physical vacuum, in other words, a "perpetual motion machine";
- vertical thermal heater with efficiency of much higher than 100%;
- solution of ecological problems (removal of pollution from rivers and bays, etc); and
- searching for oil, underground water, etc.

G. Markov *et al.*, paper about superconductivity of tungsten at 3000 K! (Rus. JTP Lett., 1996). Some time later the "scientist" Markov claimed to have invented a neutrino gun. Neutrinos were supposed to have been produced when acoustic waves passed through a vacuum! This statement characterises the level of understanding of this person. Everyone knows that acoustic waves cannot pass through a vacuum. The supposed neutrinos from this vacuum were claimed to treat cancer.

2 STRANGE PATENTS

The situation with patents in the Russian Federation changed radically in comparison with that in the Soviet Union. At present, there are no obstacles to obtaining patents for very strange and sometimes senseless methods of "treatment". The patents relate to "new medicine", "alternative medicine" as well as corresponding "devices" which have been granted official patents by the agency "ROSPATENT".

Among the "valid" patents there is a wide spectrum ranging from a "perpetual motion machine" to devices aimed at gaining money.

Several examples of officially registered "patents" are presented below.

- # 2983239 "Symptomatic treatment of illnesses with the aid of an aspen stick at the moment of a new moon for recovery of integrity of the energy shell of a person's organism".
- # 2140796 "A device for energy influences with the aid of figures at the plane generated torsion fields".
- # 2139197 "Transformation of geo-pathogenic zones to favourable ones of enormous area by the use of minerals with positive fields".
- # 2204424 "Harmonisation and improvement of a state of the biological object and surrounding space". This patent was supposed to have been utilised runic letters, which have a mystical sense according to statements of the patent authors.

The apotheosis of this stupidity is a medical device generating a gravitation field to treat patients. There are patents assigned for investigating agencies. The last patent is just one more appropriate example:

2157091 "Definition of the fact of death of a missing person by an object which belonged to him".

3 SHAMELESS ADVERTISING OF MEDICINE

There are several examples of dishonest advertising.

- Advertising of devices of "quantum medicine": "Studies in recent years, carried out in different countries have shown that hereditary information is kept not as a gene matter, but as a quantum structure. Moreover, this structure as a carrier of genetic code is arranged as a hologram... The significance of the invention of 'VITYAZ' is comparable with the flight of a man in space".
- Advertising of a pyramid: "A drink of water that has been in a pyramid for a few hours gives an absolute guarantee against oncology." In addition, pyramids are supposed to solve many other problems such as ecological issues (purification of air and water), "flu" epidemics, etc.
- "The technology of energy-informational therapy is based on fundamental properties of torsion fields... As a result of the development of energy-informational therapy technology, we can eliminate defects of the bio-field, get rid of symptoms of illness and remove the reason for sickness."

The appearance of pseudo-scientists are actually former scientists who realised that pseudo-science can bring in a much larger income than pure science.

Fortunately, most scientists remained honest people despite the terrible conditions of the 1990s.

I give several examples related to former scientists.

- Professor S. V. Zenin, a candidate of chemical sciences, doctor of biological sciences. Since 1996, he has worked in the Federal Scientific Experimental Centre of Traditional Methods of Diagnostics and Treatment (Ministry of Public Health). He claims to discriminate "real psychic from not real" by measuring changes in water acidity.
- Professor V. Voyekov (Moscow State University), doctor of biological sciences. The citation given here characterises him: "There are no doubts that various magnetic and neutron storms have an influence on water. In particular, it has been established by scientists that solar eclipses create significant changes in water properties."
- Professor K. G. Korotkov (Sankt-Petersburg State University of Informational Technology, Mechanics and Optics), doctor of technical sciences. In recent years, he changed his profession (before he studied RF discharges)

and now he claims to visualise non-existent human biofields. Here is one of his statements. "One can suppose that the fine energy of a man does not disappear with his death. This energy becomes part of a global energy field and is capable of turning into other local energy fields." Mr. Korotkov has close contact with another "scientist", Masaru Emoto, who alleges that water is capable of keeping and transferring any external information such as music, prayer, events, human thoughts and emotions!

4 STRANGE CONFERENCES

Some rather strange conferences are arranged from time to time in Russia and around the world. Here is an example of a conference held in the Barnaul State Technical University. Its official title is: "International Congress on Bioenergy Informatics".

The subject matter of the Congress was:

- Physics and technology of torsion fields;
- Bioenergo informatics in everyday life, manufacturing and in healing;
- Phenomena in the field of bioenergy informatics (anomalous phenomena, dowsing, UFO-ology, bioenergy informatics and extraordinary situations, etc.).

Another example is the conference on "Bioenergy Informational Interaction — Unity and Harmony of the World", hosted by the Moscow Technical University of Communications and Informatics.

Examples of titles of talks which were given are:

- The problem of explaining paranormal phenomena on the basis of modern theoretical physics;
- Non-electromagnetic component of electromagnetic emission — a model of physical vacuum;
- On the nature of fifth (informational) interaction.

The Russian Academy of Sciences does not take part in such conferences, but we suddenly noticed that the Institute of Oriental Studies of the RAS intended to hold an International Conference on "Ethics and Science of the Future — New Paradigm of Knowledge and Education".

It turned out that many problems which participants intended to discuss during the conference have no links with science. Here are examples of several subjects for discussion:

- Imagination of elementary particles and objects of microcosm as living structures;
- Water as a living substance;
- Reality of "channelling" phenomenon;
- New channels of communications with extraterrestrial civilizations;
- Role of geo-pathogenic zones and the study of such zones.

The most shameful fact in this story is that the Director of the Institute headed the Organising Committee. Fortunately, the conference was cancelled.

5 LARGE-SCALE SWINDLERS IN THE CORRIDORS OF STATE POWER

In the middle of the 1990s, Mr. G. Grabovoy appeared in the service of the Presidential Guard as a clairvoyant. Some time later Grabovoy read lectures on the prevention of catastrophes of extraordinary situations. In the 2001, he has published a textbook titled "Uniform System of Knowledge".

Grabovoy became widely famous after the tragic events in Beslan of 2004, when several hundred schoolchildren were killed by terrorists. He promised to bring any child back to life for a fee of $1,500.

In 2007, Grabovoy and his newspaper "Prognoz" (prognosis) predicted that Mr. Grabovoy would be the next President of the Russian Federation! Instead of that, he found himself in prison as a swindler.

A "New Leonardo" has appeared in Russia: According to the accounts of several media people from the team of this person, Mr. V. Petrik, he is an outstanding scientist who has made several discoveries of world significance. The famous programme titled "Clean Water", initiated by the speaker of the State Duma, Boris Gryzlov, was based on the "unique" filters invented by V. Petric. The cost of the Programme is estimated at USD 500 billion. Our Commission expressed serious doubts about the quality of the filters. We also mentioned that the cleaning of water by Petric's nano-filters could be dangerous for peoples' health. At the moment, the government is refusing to finance this programme.

6 SOME RESULTS OF THE ACTIVITIES OF THE COMMISSION OF THE RAS AGAINST PSEUDO-SCIENCE

1. Several communiques to the President and Government of Russian Federation were sent, each offering concrete proposals in dealing with the matter. As a result, the Commission at present receives a lot of material (including pseudo-scientific projects, patent applications, textbooks, etc.) from different government bodies for examination. Its work leads to substantial savings;
2. One of the open letters (the famous "letter of ten") to the President stopped the slide of the country towards clericalism. Children can now choose to study fundamentals of any religion, or a non-religious subject;
3. The situation with patents (in part) is improving;
4. Due to the actions of the Commission, together with the Russian Humanistic Society, licences for the "correction of biofield" were revoked;
5. Due to the actions of Commission members, the Ministry of Health revoked licences for the sale of some medicines;
6. Two fraudulent projects were stopped in Bulgaria following a request from the Bulgarian Academy of Sciences to the Commission of the RAS and our response (the papers of Russian academicians E. Alexandrov and E. Kruglyakov were published in Bulgaria);
7. Grabovoy was sent to prison;
8. The dubious "Clean Water" did not gain government support. In a new version of the project, real professionals will decide on the provision of pure water.

7 WHAT LIES AHEAD?

It is noticeable that, at present, the government is anxious at the growth and influence of pseudo-science. One of the indications of that is a request of the Security Council of Russia to the Commission. It has asked us to make proposals as to how to improve the situation. The Commission made eight proposals, including but not limited to the following:

1. To arrange that there is an obligation to involve real experts in projects that the government intends to finance;

2. To reorganise ROSPATENT for the purpose of the elimination of corruption;
3. To adopt a law on criminal answerability for dishonest medical advertising;
4. To arrange publications of cheap popular science books for youth.

If these proposals are adopted then this will make the situation better. But the problem of integrity in science can only be solved by the scientific community itself.

CHAPTER 18

THE NEED FOR GREATER ATTENTION REGARDING RESEARCH INTEGRITY IN MEXICO

José A Cuellar

Mexico is not a major research country. According to World Bank figures, it spends only about 0.45% of its gross domestic product (GDP) on scientific and engineering research. Less than 5% of its 110,000,000-strong population are engaged in some form of scientific research. Its scientists publish about 4,000 papers and register about 700 patents a year. [1] Furthermore, only about 2% of university students plan to go into research, which is leading to a lack of younger researchers to replace the current ageing researcher population [2].

The lack of support for research is accompanied by a lack of rules and regulation for the responsible conduct of research in Mexico and minimum surveillance over the proper use of funds. In this author's opinion, this situation is not a healthy one and has led to a research climate in which misconduct and other misbehaviour can flourish. We base this opinion on some troubling and unfortunate first-hand experiences that is reported here, in the hope that it will spur Mexican researchers and policy makers to take integrity in research more seriously and to develop the rules, regulations and infrastructure that will allow Mexico to become a major research country.

1 CASE STUDY

In 2007, a device created for the management of hydrocephalus provoked a series of lawsuits in different legal institutions and the Human Rights Office in Mexico. These suits alleged that the research that led to the invention of the device had been conducted without adequate permission from the Ministry of Health and without proper informed consent from the research subjects. Despite these questions, the inventor continued to publish papers about this device in international journals without any mention of the legal requirements that he had, allegedly, not fulfilled [3]. His publications on this topic stopped only after the case was finally reported in the lay press. It is currently under official investigation.

The person concerned is a high-ranking official in Mexico [4]. Before assuming this position, he gained an important international reputation as a researcher and was elected as a board member of the Mexican Agency for Science and Technology, Instead of Mexican Council of Science and Technology should say Mexican Agency for Science and Technology which administers the public funds for research activities. At the same time that he held these positions, his university received funding for a research building that included many laboratories that were under his direct supervision. In 2006, a team of researchers working under his supervision published nearly 70 articles, 21 of which listed him as an author [5]. This established this researcher as one of the most productive researchers in Mexico and, perhaps, in Latin America. Given the problems with the hydrochepalus papers and the unusually high rate of publication, the Mexican research community and government needs to review this case to assure the integrity of the work.

Further doubts about the work of this researcher and his colleagues arose in April 2008, when the main newspaper in Mexico published an article headlined "Mexican Scientist Discovered the Cause of Multiple Sclerosis" [6]. This ground breaking news was based on the results of a paper published in Annals of Neurology in March 2008 [7]. In that work, the authors reported evidence of the presence of the Varicella Zoster Virus in the cerebrospinal fluid of patients with relapses of multiple sclerosis. In an interview with the Mexican press, the editor of Annals of Neurology cautioned, "It is just too good to be true; however, we have to wait for other papers to confirm these findings" [8]. Despite his caution, the unconfirmed story was reported throughout Latin America, including on the webpage of the Mexican Agency for Science and Technology and the Office of the President of Mexico [9,10]. Unfortunately, more recent papers by recognised researchers in this field have not confirmed any of these results [11]. The details behind the publication are now being reported in the press, including a story in a popular magazine titled, "The Day When the Cause of Multiple Sclerosis was Discovered" [8].

However, the most important case occurred in the field of neurocysticercosis. The author of a paper on this topic may have claimed credit and been rewarded — by the Mexican President in one case — for work that was not his own. In January 2010, a paper was published entitled, "Neurocysticercosis: Changes after 25 years of Medical Therapy" [12]. In this work, it is stated that, "The results of the first medical treatment for neurocysticercosis were published 25 years ago", referencing the author as the original source and discoverer of the treatment. However, five years before his publication, two Mexican researchers, Manuel Chavarría (a veterinarian) and Clemente Robles (a neurosurgeon), had published the same treatment results in the medical journal Salud Pública de México (Epoca V. Volumen XXI, number 5: 603–618, September–October 1979) entitled, "Presentation of a Clinical Case of Cerebral Cysticercosis Treated Medically with a Novel Drug: Praziquantel". This paper sets out the full details for treatment, including the doses and duration of the treatment reported five years later in the later self-declared pioneering article in the New England Journal of Medicine [13, refer to reference in the paper entitled, "Therapy of Parenchymal Brain Cysticercosis with Praziquantel" (N Engl J Med 1984; 310: 1001–1007)]. The credit for the discovery of the treatment of cysticerosis should therefore be traced to Chavarría, who was a brilliant veterinarian and Professor of Parasitology at the Universidad Nacional Autónoma de México (UNAM). He was also interested in the treatment of cysticercosis and deserved credit for the discovery of its treatment, as confirmed by his colleague and co-author, Robles, a renowned neurosurgeon. In the original publication 30 years ago, Robles wrote, "The original idea of using praziquantel to treat cerebral cysticercosis was suggested to me by Dr. Manuel Chavarría, who because of his position as a veterinarian, had obvious limitations to continue his original studies in humans; then, the practical studies in humans were made by the two of us."

2 CONCLUSIONS

The outcome of this case has yet to be determined and it was reported by myself to the official authorities as the President of the Mexican College of Physicians of the National Institutes of Health (Colmexinsalud) (a non-governmental organisation). We are concerned about the integrity of Mexican research, as illustrated by this case. To push for action, we have filed two complaints: one for scientific fraud in the multiple sclerosis paper and the other for improper use of intellectual property and possible plagiarism in the medical treatment for neurocysticercosis case. The first complaint is currently under investigation, although we have not received any response to the second complaint. There has, unfortunately, been retaliation against those who signed the complaint, some of whom have been fired from their official positions.

Colmexinsalud was created to improve the environment for conducting research in Mexico, with particular attention to physicians and researchers operating within the National Institutes of Health. In particular, we are working to encourage the Mexican Congress to develop policies and standards that mirror development in other countries, as discussed at the 2nd World Conference on Research Integrity. We feel that Mexican research would benefit greatly from the establishment of a special office, an Office for the Promotion of Integrity in Research in Mexico, to promote responsible practices and to help Mexico become a country that makes significant and trustworthy contributions to the advance of knowledge. Drawing on this experience we feel that it would be easier to investigate research misconduct and encourage common approaches if the world community of researchers organised an International Commission for Research Integrity, that could join local efforts and co-ordinate multinational actions and provide resources for impartial investigations of reported cases of misconduct.

References

1. World Bank. http://data.worldbank.org/indicator/SP.POP.SCIE.RD.P6
2. Mexican Agency for Science and Technology. www.conacyt.gob-mx/SNI/paginas/default.aspx
3. Sotelo, J., Arriada, N., and Lopez, M. A. (2005). Ventriculoperitoneal shunt of continuous flow vs valvular shunt for treatment of hydrocephalus in adults. *Surgical Neurology*, 63: 197–203.
4. Comision Coordinadora de los Institutos Nacionales de Salud y Hospitales de Alta Especialidad. www.ccinshae.salud.gob.mx/interior/comisionado.htlm (October 2009)
5. Instituto Nacional de Neurologia y Neurocirugia de Mexico. www.innn.salud.gob.mx/interior/investigacion/publicaciones.htlm. (2007)
6. La Cronica newspaper (Mex). www.cronica.com.mx/nota.php?id_nota=356659. (April 12, 2008).
7. Sotelo, J., Martinez-Palomo, A., and Pineda, B. (2008). Varicella-Zoster virus in cerebral fluid at relapses of multiple sclerosis. *Ann Neurol*, 63: 303–311.

8. Revista Chilango. www.chilango.com/general/nota/2009/02/13/el-dia-que-se-curo-la-esclerosis-multiple-parte-1 (November, 2008)
9. Mexican Agency for Science and Technology. www.conacyt.mx/comunicacion/agencia/notas/salud/causaesclerosis-sotelo.htm (April, 2008)
10. Office of the Mexican Presidence. www.presidencia.gob.mx/prensa/contenido=34703 (April 2008).
11. Burgoon, M. P., Cohrs, R. J., Bennett, J. L., Anderson, S. W., Hemmer, B., Gilden, D., and Owens, G. P. (2009). Varicella Zoster virus is not a disease-relevant antigen in multiple sclerosis. *Ann Neurol*, 65: 474–479.
12. Sotelo, J. and Diaz-Olavarrieta, C. (2010). Neurocysticercosis: changes after 25 years of medical therapy. *Archives of Medical Research (Mex)*, 41: 62–63.
13. Sotelo, J., Escobedo, F., Rodriguez-Carbajal, J., Torres, B., and Rubio-Donnadieu, F. (1984). Therapy of parenchimal brain cysticercosis with Praziquantel. *N Eng J Med*, 310: 1001–1007.

SECTION IV

CODES OF CONDUCT

INTRODUCTION

Codes of conduct and other types of guidance documents play an important role in defining expectations for responsible professional behaviour in research. There are many codes of conduct to which researchers can turn for advice but these codes unfortunately vary considerably in content, utility, and availability. This is particularly true for researchers whose work is interdisciplinary and/or international.

As a first step toward understanding differences in codes of conduct and how these differences might be resolved, Melissa Anderson and Marta Shaw describe ten dimensions of codes that can impact their usefulness and potential for sharing in research collaborations. Some of the differences stem from the fact that these codes explicitly or implicitly have different goals. Some are aspirational, while others are regulatory, normative, educational and/or simply symbolic. Matthias Kaiser likewise focuses on differences, stemming from the individuals or organisations that develop particular codes, their goals, the audience they are addressing, and the relationship between the code and related legal requirements.

Anderson and Shaw see the differences in codes as a significant challenge that needs to be addressed. Some supposed normative codes do not provide good guidance for research integrity. Others have been developed without any thought of international collaboration. And some, they argue, "are often dead on arrival", either due to poor crafting or the lack of an effort to disseminate relevant information. Kaiser agrees that there are challenges but also sees some benefits in the variation, which he suggests could "trigger a lively debate in the scientific community and thus contribute to raising awareness of ethical issues." John Sulston adds to this discussion with the suggestion that codes are needed for individuals, institutions and countries and that codes must address pressing issues that impact integrity, such as excessive pressure on researchers to publish.

The remaining papers in this section describe and offer observations on specific codes. The most comprehensive in terms of coverage is the European Code of Conduct for Research Integrity described by Pieter Drenth. European-wide concern, prompted in part by a few prominent cases of misconduct, brought the European Science Foundation, the All European Academies (ALLEA) and other organisations together for the purpose of developing European-wide guidance on responsible conduct in research. Drenth summarises the key features of this code, which is available on the ESF website. Frank Wells describes a similar effort undertaken by the European Forum for Good Clinical Practice to develop comprehensive professional guidance for its members, who are mostly in the biomedical sciences.

The final papers in this section by Regnvald Kalleberg, Ashima Anand, Ping Sun, John O'Neill and Sylvia Rumball, and Timothy Dyke round out the consideration of codes with brief summaries of national experiences. The different papers should be of use to countries that are considering either adopting or revising codes of conduct for research. Regnvald Kalleberg characterises the Norwegian experience as researchers providing guidance for fellow researchers, with close co-operation with government. This bottom-up approach in a small country has resulted in codes that are fairly practical, widely used and accepted. India's efforts, described by Ashima Anand , are also largely researcher-driven but lack government support. The India Society for Scientific Values has had to proceed on its own initiative, developing guidance and pursuing individual cases in an effort to deal with misconduct in research in India. In contrast, China, as described by Ping Sun, has many codes of conduct for research that vary both in content and quality. He suggests that the development of a few good model codes might help the Chinese develop more consistent, higher quality codes. Finally, even in countries that have well-developed codes, such as New Zealand and Australia, challenges remain. John O'Neill and Sylvia Rumball describe the radical changes that are taking place in New Zealand's research universities and suggests that codes will have to be adapted to these changes. They are particularly concerned about clearly setting out obligations and duties as pressure mounts for the commercialisation of research. Timothy Dyke feels that Australian policy makers need to do more to make sure that their codes take different employment situations into account. He also mentions a problem that is common in every country and setting — finding ways to make researchers aware of their responsibility, particularly the responsibility to report and respond to integrity shortcomings by colleagues.

CHAPTER 19

A FRAMEWORK FOR EXAMINING CODES OF CONDUCT ON RESEARCH INTEGRITY

Melissa S Anderson and Marta A Shaw

The sessions on codes of conduct at the Second World Conference on Research Integrity focused on codes of conduct as formal statements of rules regarding research integrity. There are, however, few codes that focus exclusively on integrity in research. Instead, rules and expectations concerning integrity are commonly found in broader codes of conduct associated with research institutions or academic societies (Qian, 2009). These codes generally cover a wide range of topics related to proper behaviour, and they vary considerably in purpose, content, and format. They also vary in the extent to which they address research integrity.

This paper presents a general overview of codes of conduct. It presents a framework for characterising and comparing different types of codes. This framework may prove useful in the development of new codes by specifying certain dimensions that should be considered carefully and choices that should be made intentionally, not by default. The paper also discusses some of the challenges associated with codes with respect to research integrity.

1 DIMENSIONS OF CODES

The ten dimensions of codes presented here constitute a framework for characterising existing codes. The dimensions are: (1) the code's nature, (2) purpose, (3) impetus or reason for the development; (4) the subjects who must obey the code's rules; (5) its authors; (6) the extent of grounding in ethical principles or theory; (7) scope and content, (8) format, (9) language, and (10) quality. These dimensions can also serve as a checklist of choices to be made when constructing or revising a code.

The codes cited in this paper by number [#] are noted as examples that demonstrate particular characteristics. They are not necessarily exemplars of good (or bad) codes, and they do not all, in fact, specifically address research integrity.

(1) *Nature.* The English word "code" derives from the Latin "codex", meaning the trunk of a tree. In ancient times, wooden tablets were covered with wax, onto which rules and laws were inscribed (Plant, 2001). Later, the wood was replaced by sheets of paper that were bound into a codex or book, in contrast to contemporary scrolls. This derivation emphasises the written, formal nature of a code, as opposed to the implicit or informal nature of shared expectations or customs.

Codes are formal, systematic statements of rules, responsibilities and expectations of correct behaviour that have been endorsed by some sponsoring organisation. They vary substantially, however, in each of these aspects.

Some codes are as formal as legal documents. For instance, the National Ethical Guidelines for Health Research in Nepal [#30] are structured as a legislative bill and include a regulatory framework for conducting health research in the country. Other codes have a greater emphasis on guidelines or statements on good practice, as in the case of the Institute for Employment Studies Code of Professional Conduct in Socio-Economic Research, which specifically affirms that it is "primarily a guideline, or an expression of mutual understanding" [#24].

Some codes are constructed systematically on the basis of ethical principles, but most simply categorise statements by topical area. Many codes are organised around the main principles espoused by the sponsoring organisation, such as integrity, excellence, accountability, respect [University of California Statement of Ethical Values, #39] or scrupulousness, reliability, verifiability, impartiality and independence [The Netherlands Code of Conduct for Scientific Practice, #14].

A code is generally associated with a sponsoring organisation, usually an agency (governmental or otherwise), scholarly association, or research institution, and it therefore represents some form of endorsement of the rules by the organisation. Endorsement may take the form of direct notification by the organisation's leaders that the code is in effect, as in the case of universities and other research institutions. Alternatively, it may be indicated by ratification through a formal vote of the organisation's members, as often happens in academic societies. An organisation may affirm a code on nothing more than an indeterminate consensus based on a lack of strong objection by the members.

Codes can be distinguished from other statements of rules, such as laws, regulations or policies. These terms differ across countries and in translation, but in general they imply a stronger enforcement structure than is typically associated with codes.

(2) *Purpose.* Codes fall into two broad categories with respect to purpose. Most codes, one hopes, are written to improve behaviour in one way or another. Others have merely symbolic value. In many cases, codes may fulfill both purposes.

Among the behavioural codes (categorised here, in part, as in Frankel, 1989 and Moore, 2006), some focus on ideal behaviours. These aspirational codes are written to inspire people to behave as best they possibly can in their work and relationships. The American Nuclear Society's Code of Ethics [#5] makes a reference to such ideals, "Recognising the profound importance of nuclear science and technology in affecting the quality of life throughout the world, members of the American Nuclear Society (ANS) are committed to the highest ethical and professional conduct." The aspirational nature of the Singapore Medical Council's Ethical Code and Ethical Guidelines [#34] is illustrated in the following, "The medical profession has always been held in the highest esteem by the public, who look to their doctors for the relief of suffering and ailments. In modern medical practice, patients and society at large expect doctors to be responsible both to individual patients' needs as well as to the needs of the larger community. Much trust is therefore endowed upon doctors to do their best by both. This trust is contingent on the profession maintaining the highest standards of professional practice and conduct." In practical terms, ideals may not represent typical behaviour on a daily basis. Aspirational codes therefore are not intended or used as measures of appropriate behaviour.

In contrast, regulatory codes represent behavioural expectations that people must meet. They may be minimal, in the sense of setting minimal standards for behaviour, or they may go further, but in either case their purpose is to specify what people must do. Such codes are generally mandated and supported by regulatory processes by which aberrant behavior can be identified, investigated and (if confirmed as misconduct) sanctioned. Institutional codes that are regulatory in nature also include procedures for handling allegations of misconduct, with sanctions ranging from suspension or expulsion to legal action [as, for example, the Duke Medical School Honour Code of Professional Conduct, #22].

In terms of their purpose, normative codes fall somewhere between aspirational and regulatory codes, that is, somewhere between ideals and minimal standards. They represent a collective understanding of behavioural standards within a group or organisation, and their purpose is to tell members and prospective members how they need to behave if they wish to belong to the group. Those who violate the standards may be ousted. In other words, misconduct does not result in a specific punishment but expulsion from the group. A normative code is typically communicated and reinforced by socialisation and the training of newcomers, as well as repeated affirmation of and reference to the stated expectations. Among the organisations that favor normative codes are many professional associations, such as the American Educational Research Association [#2], American Psychiatric Association [#7], International Mental Health Professionals Japan [#26], and many others.

Educational codes are much like normative codes, but they are written with the explicit purpose of teaching people about the behavioural standards that they are supposed to meet. Such codes may, for example, reflect specific points that a curriculum or instructional unit should address. One example of an educational code is the Brief Introduction to Code of Ethics for Young Engineers published by the Board of Engineers of Malaysia [#17] for the purpose of training newcomers about the expectations of the professional community.

Though codes are statements about right behaviour, they sometimes have little or nothing to do with actual behaviour. Unfortunately, symbolic codes exist for other purposes. An organisation may have a code of conduct to enhance its image or reputation. Leaders may decide that a code would make the organisation appear ethical. It might be important for an organisation to signal its ethical credentials among other similar organisations, so that it can been seen as aligned with those that take ethical issues seriously and dissociated from those organisations that tolerate misbehaviour (Conzelmann and Wolf, 2005). It can serve as a point of reference in case the organisation is challenged on an ethical basis, even when most of the members do not realise that it exists and the leaders' own behaviour contradicts the code. The Enron Corporation, the subject of one of the largest accounting scandals in history, provides an example. The Enron Code of Ethics [#23] was published in July 2000, at the peak of Enron's stock prices, just one year before the corporation announced its bankruptcy. The publishing of the code could be interpreted as a legitimising measure meant to create an appearance of ethical credibility, as it was clearly contradicted by the behaviour of the company's executives. Symbolic codes may be an important component of an institution's legitimacy, enhancing its ability to persuade others to trust its actions or demonstrating its compliance with laws and regulations, though only in appearance. It is likely that every code has some symbolic significance, but most have more functional purposes aimed at affecting behaviour.

(3) *Impetus.* The impetus for a code is the initiative, directive or organisational pressure that leads to the writing of the code, not to be confused with the subjects (below). It represents the expectations that must be met in the formulation and maintenance of the code. A code may be inspired by groups external to the sponsoring organisation, such as funders, regulatory authorities, collaborating or competing organisations, or the public. In recent years, governmental legislation and threats of liability have provided the external impetus for the creation of many institutional codes worldwide. An external impetus is illustrated in the following, "The University of Alabama at Birmingham has a strong and abiding commitment to ensure that its research affairs are conducted in accordance with all applicable laws and regulations... Therefore, all UAB members ... must comply with federal and state laws and regulations applicable to any research programs" [#37]. Alternatively, a code's principal impetus may be internal, such as when a code responds to pressure from the members of the sponsoring organisation, as may happen in professional societies.

A code's original impetus may be linked to its purpose, but different combinations of purpose and impetus occur. Regulatory authority, for example, may be internal or external, and normative codes may reflect the self-expectations of professionals or the expectations of the public. The impetus for symbolic codes would naturally seem to be external audiences, but sometimes the legitimacy that they convey is most critical among the organisation's own members. As a single code may serve several purposes, so it may also respond to various impetuses. This variance can be a source of confusion and lack of clarity in a code.

(4) *Subjects.* The subjects of a code are the people who must comply with the code's rules. The subjects are generally identified by their relationship to the sponsoring organisation. An institutional code typically covers all employees of the institution [for example, Columbia University, #20 or the Barcelona Biomedical Research Park, #16]. A code of conduct for a professional society applies to the members of the society [the Anthropological Association of Ireland, #11, the American Institute of Chemical Engineers, #3]. National codes [e.g. codes of the Polish Academy of Sciences, #32, or the Swiss Academies of Arts and Sciences, #35] or regional codes [UNESCO Universal Declaration on Bioethics and Human Rights,

#36] may have less specific sets of subjects, in that they generally apply to a broader subject group. For instance, the Swiss Academies of Arts and Sciences' Integrity in Scientific Research: Principles and Procedures are addressed to "researchers, research institutions and research-promoting institutions" [#35]. Some organisations make their employees or members explicitly subject to other codes or portions of codes. For example, an academic department of a university may decide that its faculty will all be subject to the code of ethics of the primary disciplinary association, whether those faculty are members of the association or not.

Sometimes codes extend their reach beyond their natural subject groups. A code that applies to researchers in a particular field may include statements about the responsibilities of related funding or employing organisations over which it has no authority. For instance, the American Political Science Association recommends that "in making grants for research, government and non-government sponsors should openly acknowledge research support and require that the grantee indicate in any published research financed by their grants the relevant sources of financial support" [Guide to Professional Ethics in Political Science, #6]. Such statements may express the hopes or expectations of the code's authors, but they are generally non-binding on external subjects.

(5) *Authors.* The legitimacy and effectiveness of a code may be influenced by who wrote it. Codes may be written by the sponsoring organisation's leaders, members or both. In some contexts, an administratively-designed code may be the only kind that has enough authority among the subjects to be effective. In other cases, as in an academic society whose leaders are viewed as firsts among equals, the members might see a code as legitimate only if they had opportunities to shape it and ratify it by vote. This dynamic was clearly evident in the construction of the official guidelines of the Polish Academy of Sciences [#32], which involved multiple rounds of extensive feedback from participating researchers and institutions before the code could be ratified.

(6) *Grounding.* Codes of conduct are presumably grounded in some conception of good behaviour, but the extent to which this grounding is articulated in the code varies. Some codes include explicit references to ethical or other principles from philosophy, moral philosophy or science. They may even be grounded in a specific religious tradition, as the Islamic Code of Medical Ethics of the Islamic Organisation for Medical Sciences in Kuwait [#27] is rooted in the tenets of Islam. Other codes prescribe behaviour in primarily pragmatic terms, with reference to tasks and activities but little

attention to ethical theory. In the American Institute of Chemists' Code of Ethics [#4], for example, the emphasis is on chemists' duties in the conduct of their work. The American Society for Biochemistry and Molecular Biology Code of Ethics [#9] translates the general goal of "advancing human welfare" into specific obligations of researchers towards the public, other investigators, and trainees.

The titles of codes do not reliably distinguish between those with and without explicit grounding in ethical theory. Some "codes of ethics" are less explicitly grounded in ethical principles than some "codes of conduct", which one might assume to be more pragmatically oriented. The Professional Ethical Guidelines for Biomedical Laboratory Scientists of the Norwegian Institute of Biomedical Science are indeed as concise as the term "guidelines" might suggest, but they are supplemented in the booklet Ethics for Biomedical Laboratory Scientists [#31] with summaries of meta-ethics, descriptive ethics and normative ethics. In contrast, the Ethical Principles of the Association for Applied Psychophysiology and Biofeedback [#12], whose title suggests a close link to ethical theory, actually consist largely of behavioural directives. Some codes, such as the one published by the University of California, San Francisco [#38], use the terms "code of conduct" and "code of ethics" interchangeably.

(7) *Scope.* Codes of conduct vary widely in scope and content (see Iverson *et al.*, 2003, and Jones, 2007, with regard to research integrity content). Some codes cover only a few points [ten in the case of the Ethical Principles of the Association for the Study of Higher Education, #13], from which the code's subjects will presumably infer guidance on other matters. By contrast, other codes present rules on everything from overall behaviour to specific interactions among researchers. This difference in scope can be seen even within a single disciplinary field: the Guidelines for Good Professional Conduct of the Political Studies Association in the UK [#33] are only 2.5 pages in length, covering essential topics related to the profession, whereas the American Political Science Association's Guide to Professional Ethics in Political Science [#6] is 38 pages long and includes principles as well as their application in many professional contexts, with references to specific situations and expected behaviour.

(8) *Format.* Given the substantial variation in scope and content of codes, it is not surprising that codes also differ widely in format. They may be detailed or sketchy, succinct or lengthy, bulleted or discursive, well-organized or jumbled. Some include a brief history of the code-construction process, especially in cases when it involved a broad collaborative endeavour

or was enabled by external assistance [National Ethical Guidelines for Health Research in Nepal, #30, National Code of Health Research Ethics in Nigeria, #29].

Other components that appear in some codes are as follows, with examples: definitions ["'Moral rights' means fundamental and inalienable human rights that might or might not be fully protected by existing laws and statutes", Canadian Psychological Association, #19]; principles ["Respect, Competence, Responsibility, Integrity", British Psychological Society Code of Ethics and Conduct, #18]; statements of good or best practice ["Every scientist has to document in writing enough essential information on the execution of an experiment to enable independent experts to repeat it", Medical University of Graz, Standards of Good Scientific Practice, #28]; prescriptions ["All members of the International Association for Dental Research shall communicate in an honest and responsible manner", International Association for Dental Research Code of Ethics, #25]; and guidelines specific to students or other newcomers ["Also, beginning researchers have a responsibility to seek out and work with mentors rather than expect that potential mentors will seek them out", On Being a Scientist #21].

(9) *Language.* The language of codes is almost universally prescriptive, indicating required, desired or forbidden behaviours. Accordingly, most are worded in terms of what the subjects must do ["The (Biomedical Laboratory Scientist) must demonstrate respect for life and for the inherent dignity of human beings", Professional Ethical Guidelines for Biomedical Laboratory Scientists of the Norwegian Institute of Biomedical Science, #31] or should do ["Educational researchers should attempt to report their findings to all relevant stakeholders", Ethical Standards of the American Educational Research Association, #2]. In the English version of the Bahrain Society of Engineers' Code of Ethics [#15], each point notes what engineers shall do, "Engineers shall avoid deceptive acts and shall not abuse public or private office for personal gain." The imperative is even more explicit in the Ethical Guidelines for Statistical Practice of the American Statistical Association, as in the following excerpts, "Strive for relevance in statistical analyses... Use only statistical methodologies suitable to the data and to obtaining valid results... Disclose conflicts of interest, financial and otherwise, and resolve them" [#10].

A few codes, however, refer to what subjects actually do, as in the following: "Psychologists seek to promote accuracy, honesty, and truthfulness in the science, teaching, and practice of psychology" [Ethical Principles for Psychologists and Code of Ethics of the American Psychological Association, #8]. The Code of Ethics of the American Association of

Pharmaceutical Scientists [#1] claims that "In their scientific pursuits, (AAPS member scientists) conduct their work in a manner that adheres to the highest principles of scientific research." With this language, which aligns with the purposes of normative codes, organisations specify the behaviour that is expected as a condition of membership in the organisation; those who do not behave according to the code may not continue as members.

(10) *Quality.* Good codes — like good policies or regulations — are clear, well-written, and easy to understand, interpret and use. They provide adequate guidance on the most important issues that come up in the work of the subjects, and they are general and adaptable enough to be of use as new situations arise.

The previous descriptions of code dimensions have emphasised variation. Codes exhibit these variations because of significant differences in organisational context and purpose. A code's purpose, scope, format and so on need to be aligned with the needs of the organisation. That is not to say, however, that codes should comprehensively and expansively represent all possibilities related to these dimensions. On the contrary, codes should be designed with these dimensions in mind and with explicit decisions on each dimension. Good codes are written with a specific purpose, for specific subjects, by authors whose contributions enhance the codes' legitimacy and effectiveness, and with deliberate attention to the appropriate grounding, scope, format and language. There may be more than one purpose or impetus for a good code, as long as such multiplicities are identified or incorporated appropriately. A bad code, by contrast, is rather a mess. It may be a mix of ideals and minimal standards, a jumble of rules and purely symbolic statements, all written to satisfy multiple constituencies or control the behaviour of disparate groups of people.

The quality of a code is a subjective characteristic. Its actual effectiveness may be impossible to measure. In fact, there is very little evidence on what, if anything, makes a code effective in meeting its purposes, whether behavioural or symbolic.

2 CHALLENGES ASSOCIATED WITH RESEARCH INTEGRITY IN CODES OF CONDUCT

There are three broad challenges associated with the ways research integrity is currently addressed in codes of conduct. None of the challenges has a ready solution, but each can be mitigated by responses suggested here.

Challenge 1: Inadequacy of Codes of Conduct in Addressing Research Integrity. Research integrity is typically incorporated into codes of conduct that have a much broader scope. It may appear under headings such as "misconduct" or "plagiarism", or it may be identified in phrases scattered throughout the document. Some codes include a general admonition to maintain the highest standards of integrity, whereas others address it only in the negative, through rules against fabrication and falsification. There are research institutions and organisations that simply do not include research integrity in their codes of conduct, and, remarkably, there are others — like the American Economic Association and the (U.S.) National Economic Association — that have no code of conduct at all.

Even codes that address research integrity can be ineffective. Some are poorly written or confusing. Some leave open opportunities to get around rules. Others do not offer guidance on common ethical dilemmas that researchers face in the work context. Many have no meaningful sanctions associated with violations of the rules.

Response. A useful response to this challenge is to build on good examples of existing codes. Groups that have the responsibility for writing or revising a code can benefit greatly from examining other codes in similar institutions and contexts, instead of starting from scratch. Some of the codes discussed at the Second World Conference on Research Integrity and described in this volume provide useful examples. The hundreds of codes available online provide a wealth of options from which good choices can be made. Valuable online resources include the following:

- UNESCO Global Ethics Observatory Database (www.unesco.org/new/en/social-and-human-sciences/themes/ global-ethics-observatory/access-geobs) allows the user to search a global sample of codes by their features as well as geographical coverage;
- Virginia Polytechnic Institute and State University Database of Codes of Conduct from Around the World (courses.cs.vt.edu/cs3604/lib/WorldCodes/WorldCodes.html) contains a large number of codes from various regions of the world, as well as recommendations for developing new codes;
- Codex Database (www.codex.vr.se/en/etik9.shtml) provides guidance to researchers on conducting research in accordance with local codes and laws, and also includes educational materials and weekly updates on research ethics;

- National Institute of Engineering Online Ethics Centre for Engineering and Research (www.onlineethics.org) contains a large database of codes, as well as educational materials and annotated bibliographies on related topics.

Though individuals and committees who are working on codes should consult existing documents, it is important that the new codes acknowledge all sources of language and ideas, to avoid plagiarism. Ironically, instances of identical or similar language in the many codes we reviewed suggest that this is not always done.

Challenge 2: International Research Collaboration. There is at present no global code of conduct for research. Researchers at a particular research institution may be covered by a code of conduct. Researchers at other institutions are covered, if at all, by different institutional codes of conduct. National codes extend only to national boundaries, and regional codes are similarly limited. Disciplinary codes apply only to their own disciplines and, as a practical matter, only to researchers who are members of the sponsoring disciplinary society, which may also have national boundaries. As noted above, even where codes exist, they may not include research integrity or provide adequate guidance on research issues. In short, an individual researcher is likely to be subject to a patchwork of codes, at different levels and with different aims.

The challenge here is that institutional, national, regional and disciplinary codes do not match the reality of how science is increasingly done: through international research collaborations. These collaborations are typically international, interinstitutional and interdisciplinary, all at the same time. Each member of the collaboration is subject to a different set of codes (as well as laws, regulations and policies), and they are not likely to work through all the disparities unless or until problems or violations arise.

Response. There is a need for agreement on some basic principles of responsible research conduct. The Singapore Statement prepared for and discussed at the Second World Conference on Research Integrity provides a starting point for this kind of agreement.

Challenge 3: Dead Codes. Unfortunately, codes are often dead on arrival. It is as though, once completed, they are written on the sections of a dead tree trunk or bound into a codex book and buried away. They are forgotten or ignored. The best codes in the world are worth nothing if they are dead.

Response. Codes need to be treated as living documents. They should be updated and revised on a regular basis, as this practice attracts renewed attention (Frankel, 1989). They should be taught to all young researchers — not just handed to them to read, but discussed at length. Junior researchers need to consider what the rules mean, how they have been interpreted in the past or how they might be interpreted in new circumstances, and what limitations or caveats apply. Those who produce codes of conduct may have clear ideas about their intent, but working scientists — especially newcomers — may misinterpret the rules in a given context or fail to see the relevance of a general standard to a specific research situation. Ongoing discussion and debate will bring some of these points of confusion into the open to be addressed.

Most importantly, codes should become an integral part of the work context. When codes are confined to initial training or orientation sessions, they are quickly forgotten. Instead, senior researchers should deliberately make references to codes, rules and standards of research integrity on a regular basis, so that ethical considerations become part of the natural pattern of scientific work in the laboratory or other research settings.

Codes will not make anyone behave well. If they are maintained as living documents, however, they may lead people to consider their everyday work in the light of ethical principles and standards of good conduct. As such, they can serve an important role in helping people to make good behavioural decisions that will ensure the trustworthiness of research, which is the essence of research integrity.

3 CODES

(All codes listed here were accessed on 30 June, 2010)

1. American Association of Pharmaceutical Scientists. *Code of ethics.* Accessed from www.aapspharmaceutica.com/inside/refguide/CodeofEthics.pdf
2. American Educational Research Association. *Ethical standards of the American Educational Research Association.* Accessed from www.aera.net/uploadedFiles/About_AERA/Ethical_Standards/EthicalStandards.pdf
3. American Institute of Chemical Engineers. *Code of ethics.* Accessed from http://www.aiche.org/About/Code.aspx
4. American Institute of Chemists. *Code of ethics.* Accessed from http://www.theaic.org/DesktopDefault.aspx?tabid=46

5. American Nuclear Society. *Code of ethics.* Accessed from http://www.ans.org/about/coe/
6. American Political Science Association. *A guide to professional ethics in political science.* Accessed from http://www.apsanet.org/content_9350.cfm
7. American Psychiatric Association. *The principles of medical ethics with annotations especially applicable to psychiatry.* Accessed from http://www.psych.org/MainMenu/PsychiatricPractice/Ethics/ResourcesStandards.aspx
8. American Psychological Association. *Ethical principles of psychologists and code of conduct.* Accessed from http://www.apa.org/ethics/code/index.aspx
9. American Society for Biochemistry and Molecular Biology. *Code of ethics.* Accessed from http://www.asbmb.org/Page.aspx?id=70&terms=ethics
10. American Statistical Association. *Ethical guidelines for statistical practice.* Accessed from http://www.amstat.org/committees/ethics/index.cfm
11. Anthropological Association of Ireland. *Ethical guidelines.* Accessed from http://anthropologyireland.org/research1.htm
12. Association for Applied Psychophysiology and Biofeedback. *Ethical principles.* Accessed from http://www.aapb.org/ethical_principles.html
13. Association for the Study of Higher Education. *Principles of ethical conduct.* Accessed from http://www.ashe.ws/?page=180
14. Association of Universities in the Netherlands. *The Netherlands code of conduct for scientific practice.* Accessed from http://www.vsnu.nl/web/show/id=75803/langid=42/contentid=1095
15. Bahrain Society of Engineers. *Code of ethics.* Accessed from http://www.mohandis.org/en-US/Code_Of_Ethics.aspx
16. Barcelona Biomedical Research Park. *Code of good scientific practice.* Accessed from http://www.prbb.org/eng/part01/p06.htm
17. Board of Engineers Malaysia. *Brief introduction to code of ethics for young engineers.* Accessed from http://www.bem.org.my/circulars/dosdonts.pdf
18. British Psychological Society. *Code of ethics and conduct.* Accessed from http://www.bps.org.uk/document-download-area/document-download$.cfm?file_uuid=E6917759-9799-434A-F313-9C35698E1864&ext=pdf

19. Canadian Psychological Association. *Canadian code of ethics for psychologists.* Accessed from http://www.cpa.ca/cpasite/userfiles/Documents/Canadian%20Code%20of%20Ethics%20for%20Psycho.pdf
20. Columbia University. *Administrative code of conduct.* Accessed from http://www.columbia.edu/cu/compliance/pdfs/Code_of_Conduct.pdf
21. Committee on Science, Engineering, and Public Policy, National Academy of Sciences, National Academy of Engineering, and Institute of Medicine (US). (2009). *On being a scientist: A guide to responsible conduct in research.* Washington, D.C: National Academies Press.
22. Duke University School of Medicine. *Honor code of professional conduct.* Accessed from http://www.medschool.duke.edu/wysiwyg/.../SOM_Honor_Code_07_08.doc
23. Enron Corporation. *Code of ethics.* Accessed from http://www.thesmokinggun.com/documents/crime/enrons-code-ethics
24. Institute for Employment Studies (UK). *Code of professional conduct in socioeconomic research.* Accessed from http://www.respectproject.org/code/respect_code.pdf
25. International Association for Dental Research. *Code of ethics.* Accessed from http://www.dentalresearch.org/i4a/pages/index.cfm?pageid= 3562
26. International Mental Health Professionals Japan. *Code of ethics.* Accessed from http://www.imhpj.org/about/code_of_ethics.html
27. Islamic Organisation for Medical Sciences (Kuwait). *Islamic code of medical ethics.* Accessed from http://www.islamset.com/ethics/code/index.html
28. Medical University of Graz. *Standards of good scientific practice.* Accessed from http://www.meduni-graz.at/images/content/file/forschung/gsp/GSP_Standards.pdf
29. National Health Research Ethics Committee of Nigeria. *National code of health research ethics.* Accessed from http://www.nhrec.net/nhrec/code.html
30. Nepal Health Research Council. *National ethical guidelines for health research in Nepal.* Accessed from http://www.nhrc.org.np/guidelines/index.php
31. Norwegian Institute of Biomedical Science. *Ethics for biomedical laboratory scientists.* Accessed from http://www.nito.no/dm/public/115266.PDF
32. Polish Academy of Sciences. *Good manners in science: A set of principles and guidelines.* Accessed from http://www.ken.pan.pl/images/stories/pliki/goodmanners.pdf

33. Political Studies Association (UK). *Guidelines for good professional conduct.* Accessed from http://www.psa.ac.uk/Content.aspx?ParentID=3
34. Singapore Medical Council. *Ethical code and ethical guidelines.* Accessed from http://www.smc.gov.sg/html/1153709460047.html
35. Swiss Academies of Arts and Sciences. *Integrity in scientific research: Principles and procedures.* Accessed from http://www.swiss-academies.ch/downloads/Layout_Integritaet_e_online_000.pdf
36. UNESCO. *Universal declaration on bioethics and human rights.* Accessed from http://portal.unesco.org/en/ev.php-URL_ID=31058&URL_DO=DO_TOPIC&URL_SECTION=201.html
37. University of Alabama at Birmingham. *Research code of conduct.* Accessed from http://main.uab.edu/show.asp?durki=63288
38. University of California San Francisco. *Campus code of conduct.* Accessed from http://ucsfhr.ucsf.edu/files/finalcc.pdf
39. University of California. *Statement of ethical values.* Accessed from http://compliance.uclahealth.org/workfiles/PDFs/UC_Standards.pdf

References

1. Conzelmann, T., and Wolf, K. D. (2005). The potential and limits of governance by private codes of conduct. In *Transnational Private Governance and its Limits.* Graz, J. C. and Nölke, A. (eds.), London: Routlege, pp. 98–114.
2. Frankel, M. S. (1989). Professional codes: Why, how, and with what impact? *Journal of Business Ethics*, 8, 109–115.
3. Iverson, M., Frankel, M. S., and Siang, S. (2003). Scientific societies and research integrity: What are they doing and how well are they doing it? *Science & Engineering Ethics*, 9(2), 141–158.
4. Jones, N. L. (2007). A code of ethics for the life sciences. *Science & Engineering Ethics*, 13(1), 25–43.
5. Moore, G. (2006). Managing ethics in higher education: Implementing a code or embedding virtue? *Business Ethics: A European Review*, 15(4), 407–418.
6. Plant, J. F. (2001). Codes of ethics. *Handbook of Administrative Ethics.* Cooper, T. L. (ed.), New York: Marcel Dekker, pp. 309–333.
7. Qian, M., Gao, J., Yao, P., and Rodriguez, M. A. (2009). Professional ethical issues and the development of professional ethical standards in counseling and clinical psychology in China. *Ethics & Behavior*, 19(4), 290–309.

CHAPTER 20

DILEMMAS FOR ETHICAL GUIDELINES FOR THE SCIENCES

Matthias Kaiser

I shall address the question of what considerations one is confronted with when designing ethical guidelines for science, and I am going to discuss four sub-questions in particular. The basis for my reflections would be my experiences when I was centrally involved in the drafting of the Guidelines for Research Ethics in Science and Technology of the Norwegian National Committee for Research Ethics in Science and Technology (NENT). Another source of inspiration was the International Council for Science (ICSU), Standing Committee on Responsibility and Ethics in Science study on the ethical guidelines for science, in which I was also centrally involved. Thus the following deliberations should not be viewed as the result of a systematic study, but as the outcome of personal experience.

I claim that four principal dilemmas will typically occur whenever one has to make decisions that will affect the content and format of the guidelines one is drawing up. The dilemmas are as follows:

1. The first one is the question of identification, i.e. how does one conceive that an individual scientist perceives the guidelines as binding?
2. The second one is the question of the goal of setting down guidelines, i.e. are they intended to spell out what in fact are current norms of scientific practice, or are they intended to actually move the current practices of science closer to an ideal ethical norm?
3. The third one is the question of the intended audience of the guidelines, i.e. are they directed mainly or even exclusively towards practicing scientists, or are they, in addition to the former, also directed towards the general public that takes an interest in science?

4. The fourth one is the question of the relationship to legal measures, i.e. are they rules for self-policing within the scientific community, or are they instruments to protect parties that are affected by scientific research?

I shall address these four questions in turn.

1 IDENTIFICATION

If guidelines are to make a difference at all, then they need to be perceived as binding or rule-giving by someone, typically by practising scientists and researchers. It is well known in sociology that common norms and values are the expression of the cohesion that binds a social group or institution together. Thus, an individual needs to identify with the social group in question in order to feel its norms as personally binding. Now, a practising scientist is typically aligned to a number of important group identities. First and foremost, a scientist is a member of a disciplinary group, the field in which one is trained and in which one has achieved the merits to perform research. This identity refers to a large international community of researchers with common traditions and their own channels of communication such as specific scientific journals. They are typically organised in scientific unions. This group in turn is a part of an even larger international community of all sciences, typically organised in umbrella organizations like the ICSU, and often represented in national or regional academies. This overarching identity reflects the fact that all science shares some interests, and, at least in the eyes of the public, they all engage in the study and analysis of reality. Often there are also additional identities cross-cutting the above. This is, for instance, the case in all problem- or topic-related research like environmental studies, climate research, cancer research and the like. In these cases the field of interest defines an additional identity.

However, parallel to the above there are obviously also institutional identities. Scientists need to have a work place and they are employed by institutions that perform research, such as universities or institutes. Typically these institutions will employ researchers with varying disciplinary or other identities. It is also these institutions that bear a certain responsibility for the quality of the work that goes out from them. The institutions will in most cases have a local or regional profile. They will also be subject to national regulations and laws pertaining to the conduct of research within a country.

These are but a few of the most relevant professional identities a scientist may be subjected to. Both horizontally and vertically, they operate at different levels and respond to different challenges. The question is whether these different identities imply a possible conflict in relation to the issue of adherence to guidelines.

It is obvious that when codes or guidelines are developed, research institutions, for example, will address different issues (such as teaching obligations or administration) than a disciplinary union. In other words, one set of guidelines cannot replace another. It is furthermore a striking fact that there seems to be a kind of competition between guideline-emitting institutions. Guidelines seem to be seen as a kind of authority that speaks on behalf of science or at least on behalf of a section of science, and as such they can become a strategic instrument to raise the perceived status of the institution. The diminishing role of academies in several European countries is seemingly countered by their efforts to be authoritative on normative issues in science. Thus, they may be in direct competition with funding agencies, national ethics councils or other such bodies.

What are the implications of this for the design of ethical guidelines for science? Personally, I believe that the implications are only very serious if one adheres to an outdated model of the unity of science. Such a way of thinking would imply a hierarchy of authority, based on unique fundamental epistemic norms for all research, which is in stark contrast to the multiplicity of actors who claim to voice valid scientific norms. However, if one supports a more pluralistic view of scientific methods and practices, including their normative dimensions, then this situation need not be so problematic. Since I subscribe to the latter, I welcome the issuance of ethical guidelines from various actors, even if they do not overlap in each and every detail. They may all still be very fruitful as constructive tools to raise ethical awareness among scientists. Problems arise only if specific rules directly contradict each other. If one set of guidelines would, for example, condemn honorary authorship while another finds it acceptable, then one has a problem. Yet, this seems seldom the case, and if such contradictions are detected, they typically spur a lively discussion over how best to resolve the issue.

My conclusion in regard to this dilemma is that it is vitally important that one reflects thoroughly on the remit and task of the body that is to maintain the guidelines, and that this is worked into the specific profile they will have in the end.

2 GOALS

Why is it that one feels the need to spell out ethical guidelines for science? Obviously one implicitly admits that some problems do occur in the practice of research and science: if everything were perfect, then one would not need to formulate guidelines. But what is the nature of these problems? With regard to the latter question, people are assumed to be of different minds. Some people feel that within the scientific community there are morally weak characters, "bad apples" so to speak, as one would expect in every human practice and profession. In this case, guidelines are meant to support the large majority who do already follow those rules in their effort to correct the bad apples. However, some other people assume that breaches of good ethics are the result of systemic factors pertaining to the conduct of science, and that it is due to conflicting pressures that problems occur. Such conflicts may be related to the epistemic role of science finding truths versus the role of science as supplying commercially valuable commodities in a knowledge market. In this latter case, the goal of guidelines would serve to clarify priorities among goals and to move science closer to its genuine goals.

These differences have implications for how one thinks such guidelines can be justified. In the first case, one aims at spelling out what a large majority of scientists already accept. Thus, a kind of widespread consensus would provide the justification for the guidelines. If in doubt, one can always support the guidelines empirically with evidence of actual scientific practices. In the second case, one cannot expect this kind of consensus, but one addresses issues that are debated in the scientific community and takes a normative stance on them. In other words, the content of the guidelines needs to be justified by reference to a theory of basic values of science. Thus, it is by reference to what scientists should do, rather than what they actually do, that support is given for the guidelines. Obviously, such a position is much more controversial than the first one. One enters the realm of ethics proper and this is typically seen to be a wide field.

In practice, I believe that many existing guidelines are a mixture of both positions. However, I also think that it is useful to reflect on these issues right at the start, since they will have consequences for how ambitious the guidelines in effect will be and how much resistance they will trigger. Is one to stick one's head out and take on normative opposition, or is one to seek a smooth integration into the workings of science? I personally prefer to be rather highly ambitious with the guidelines and thus create a lively debate. It is through debate that learning occurs, and I believe in the field of the ethics of science we still have a lot to learn. However, others may think otherwise. In any case, clarification of the goals will be useful.

3 AUDIENCE

Who is the intended audience of the guidelines? Obviously one wants scientists to be the primary audience of the guidelines, especially young scientists. If they do not take note of the guidelines, much of the effort will have been wasted. However, scientists do not need to be the only audience of the guidelines. Ethical guidelines may equally address the general public as an audience, as part of the science-society relationship. This is known from business life where rules of corporate social responsibility are also communication tools with stakeholders and the public. Similarly, ethical guidelines for science may function as a communication tool with society. They may be something akin to a "showcase" of science. We find references to this function in many background notes on such guidelines. Often it is claimed that science needs to restore public trust in science (which is conceived as lacking these days), and this is to be achieved by clarifying the values of scientific inquiry. Since parts of society supposedly question the important roles of science, and perhaps more importantly the value-bias they detect in it, it is the role of guidelines to assure society that science is still on the right track and that there is harmony between social values and scientific values. But there may be other social functions as well. In some sense, guidelines may also be viewed as advertising the advantages of science in a market of competing sources of knowledge. This is perhaps the case with regards to science as a policy advisor where other sources, like NGOs, are viewed as having taken on roles that previously were the exclusive domain of science. Guidelines may also be advertising science in a different manner: there is a problem recruiting sufficient numbers of young people to a scientific career in many rich Northern countries. Guidelines may appeal to young people in particular because they focus on the basic function of science that is sometimes lost in its public image.

Yet, guidelines may also have a more restricted function. They can simply be viewed as a communication tool within the scientific community. They can exclude certain forms of practice that seem to be spreading, such as the sloppy use of data, and they can thus be used as a tool to discipline those who deviate too far from common practice. They can also be a source-text in the education of young scientists, since the transmission of values through direct teacher-student interaction may be threatened in fields with large numbers of students. Finally they may also clarify ongoing disputes about good practice within the scientific community.

Again, in practice, many guidelines may intend to reach both audiences. However, it does have consequences in terms of actual content if one also explicitly aims to communicate with society with the help of ethical guidelines. In this case, it is vitally important that the guidelines relate to overarching social values that society readily accepts as virtuous features. Respect for human rights and aiming at sustainable development may be a case in point. Simply spelling out internal values of the scientific community may not be sufficient to clarify how these values actually relate to external societal values. After all, it is with reference to the external values that society is willing to invest a part of its funds in public research. Naturally, the more comprehensive and specific that one is about these external values, the more one may be subjected to criticism, since these external, societal values typically exist in the realm of politics as well. There is no general recipe on how to address this, but again my feeling would be that it is better that the guidelines are debated and criticised by some, than not being discussed at all.

4 LAW

Scientific research and scientific institutions enjoy a kind of autonomy that is unparalleled in public life. There is no high court for good science, no final authority on truth. It is the scientific community itself that validates knowledge claims. The scientific community also self-polices in the event of violations of its most basic rules. And science is universal — national or political boundaries are not involved or instrumental in the workings of science, implying among others privileges the freedom of communication between scientists (see, for example, ICSU's Blue Book). All this is also referred to as the freedom of science. This freedom and autonomy of science is in some countries even guaranteed in their constitutions, such as in Germany. The effect is a marked restraint on lawmakers and public authorities when it comes to interfering with the workings of science. To a large extent, this is very beneficial for science. It has, among others, the advantage that this tends to stimulate ethical reflection within science itself. People that meet to discuss the ethics of science are typically scientists themselves. It has also brought about what I consider a healthy inclination to avoid red tape in the workings of science.

Yet, this is only one side of the coin. The other side of freedom is, of course, responsibility. There are indeed things related to research for which one has to take responsibility. This applies specifically to what may be described as the protection of weak parties that are affected by research. One prominent example is research involving human subjects. As a result of unethical medical experiments with human subjects, existence of which emerged during the Nuremberg trials, the world has agreed to set down binding international standards. They are enforced by law; if you breach them, you can be punished by law. Another example is animal research, which also is regulated by law in most countries. One could add other examples as well, such as intellectual property rights, data protection and privacy. The protection of whistleblowers is currently much discussed and they are already offered legal protection in many countries.

In sum, there is a tender relation between legal measures and the internal ethics of science. How is this to be reflected in ethical guidelines for science? Are ethical guidelines but another measure of protection alongside legal measures? Are they law-like in their binding character?

Ethical guidelines are an example of what often is termed soft law (as opposed to hard law). One may wonder how it is that soft law has an effect at all, if it is apparently not enforced with the same rigour as hard law. Are ethical guidelines simply nice words, perhaps window dressing? I do not think they necessarily are. A study of the introduction of internal control at Norwegian oil platforms convinced me that they can make a difference. Internal control means that the lawmaker only provides a framework law that specifies that the operator, in this case an oil company, has to work out guidelines for health, safety and environmental protection in the workplace. It does not specify what these measures actually are. The point is that the introduction of internal control actually spurred a number of positive changes in the operations of these companies, among others changes of management, such as reporting incidents as the first item on every management agenda.

I view ethical guidelines as fulfilling a similar function. They will need to be supported by some framework laws in each country. The nature of these laws will depend on the country in question, since research and science is regulated differently in each country. Yet, ethical guidelines should have the potential to make a clear difference in the conduct, management and education of science. Often this will imply at least some additional administrative burdens, such as, for example, permissions from competent ethical bodies, but the extent and nature of this can vary and be worked out in co-operation with the scientific community. It will also imply a recognised standing in the education of scientists, but the nature and extent of it may well also vary as scientists see fit. In the end, they will still be accountable for the kind and effectiveness of the practices they endorse.

Thus, I see no principal conflict between law and ethics. The challenge is to strike the right balance between them and to understand the specific role each can play in the workings of science.

5 CONCLUSION

I have discussed four principal issues that may arise when designing ethical guidelines for science. One needs to reflect on these questions, but there is no unique or best answer to any of them. Consequently, there is no ideal format for ethical guidelines in science either. There is room for a variety of different guidelines, with different formats and different functions. The hope is that they trigger a lively debate in the scientific community and thus contribute to raise awareness of ethical issues. In that way, they may perhaps not solve all ethical problems in science, but they may at least be useful tools for the resolution of some pressing problems.

CHAPTER 21

LEVELS OF RESPONSIBILITY[1]

John Sulston

Trust between individuals is a basic requirement for the conduct of science. Obviously enough, this requires a balance. On the one hand, in order to get anything done we need a measure of personal ambition — a determination to solve problems and to be excited by the challenges in doing so. On the otherhand, we need to share our results. Both those aspects are important, and to get the balance right requires personal integrity. The key at the personal level is not to get so bound up with our ambition that we cannot let anything go, or that we are tempted to steal from others.

Plagiarism is always hurtful, but so long as we have enough projects going on and we are excited about the whole process, that is usually a minor matter compared with the much more rewarding aspects of discovering and sharing and gaining peer approval — because unless we are very unusual (humans being social animals), we are very much motivated by the encouragement of our colleagues.

So, along with personal integrity goes collective integrity. As well as providing support, collective action is an attribute of self-policing science. It does not have to be nasty — just a matter of noticing what is going on. In a lab, or in a group, you become aware if someone is cheating, and the best way is to nip it in the bud by being willing to question them. It is difficult, of course, if the person involved is senior to you. We must recognise that when somebody junior does have to make criticisms, they are stepping out of line in a heroic fashion. Not everybody is willing to do

[1]Adapted from: Sulston, J. (2008). Science and ethics. *The Biochemist*, 30(6): 4–6.

that, and therefore, it is very important that we support whistleblowers, and if someone is *genuinely* revealing malpractice in a lab, then they must be protected from victimisation.

More commonly, though, misconduct occurs at the junior level but is not recognised by the senior scientist. When that happens, it is a clear failure of the supervisory process. This may be nobody's fault, but there is one way in which the structure of science contributes to the problem — namely the trend towards ever more publications each year. A practical proposal, which we have been discussing in CFRS, is to discourage scientists from publishing so many papers. At every university there are examples of people who have their name on more papers per year than they can possibly fully comprehend. A person should not have their name on a paper without really knowing its contents and what went into it; if they are the head of the laboratory, they should not automatically be author but should be acknowledged for the administrative role that they fulfil. Full authorship should be reserved for papers where they really have made a contribution.

If we can begin to reduce this pressure for more and more publications, by convincing search, promotion, funding and award committees that an overly long publication list should attract a negative mark, then we shall have done science a service. A further benefit of such a change would be the reduction of pressure on reviewers, so that they would have time to investigate manuscripts more fully and would have a better chance of detecting malpractice.

This takes us to the level of institutional integrity, which is very important and not sufficiently discussed. It comes in many forms. Most scientists work in companies, large numbers work for governments, and others in universities. A company owes a duty to its shareholders: a commercially promising drug in the pipeline can easily lead to pressure not to disclose adverse effects (for example, GlaxoSmithKline's pressure on John Buse over Avandia) or to set up trials in such a way as to provide a favourable outcome. Much overt malpractice in this field has been exposed and reduced, but more subtle forms continue unabated — for example misrepresentation in marketing (for which Pfizer was fined $2.3bn last year). Universities, too, are not immune from these pressures — either because of corporate links (think of Nancy Olivieri and others at Toronto), or because they are competitive businesses in their own right. They could play a key role in reducing the pressure for publication. Government scientists can experience very substantial conflict. We know, for example, that under the Bush administration in the USA, people in NASA were prevented from speaking out about climate change. We know that in the UK one scientist (David Kelly) was actually hounded to his death because he blew the whistle on fake information in the dossier that supposedly justified the Iraq war.

Extending from the institutional to the national and policy level, we sadly find much indirect incitement to misconduct. For example, explicit payments by some developing countries for papers in *Nature* are clearly a danger, and so is the expressed view of some senior scientists in developed countries that financial greed is good.

Beyond misconduct, there are reasons for concern at all levels about the possibility of malicious acts. Constraints are not straightforward, because potentially harmful technology is mostly dual use — that is to say we need it to continue making discoveries, inventions and useful products. Probably the best safeguards lie in personal and collective integrity — the eyes and ears of the scientific community. It is up to us all as scientists to be guardians of institutional integrity. For this to work, there must be strong policies that help people not only to detect malpractice but to act effectively when they find it.

In conclusion, codes of conduct need to operate at all levels, and indeed, most strongly at the top. It is not enough to simply provide guidance to individuals, and subsequently castigate them if they behave unethically. We must ask equally about the pressures that led them to behave in that way, and be aware that we and our institutions may have created an environment in which they had little choice but to do so.

CHAPTER 22

THE EUROPEAN CODE OF CONDUCT FOR RESEARCH INTEGRITY[1]

Pieter J D Drenth

Codes of Conduct have been developed to help researchers understand their responsibilities. Such Codes have been developed by universities, research institutes, professional organisations, academies, national governments, or even supranational institutions. They differ significantly in purpose, content, scope and authority. The sessions within track 2 explore the different types of codes, their purposes, their strengths and weaknesses, and the experiences in implementing them. Examples of Codes are passed in review and provide an opportunity for comparative analysis of the different types in existence. An example of a supranational Code is the "European Code of Conduct for Research Integrity".

1 HISTORY

European research institutions, academies and professional organisations became concerned and wanted to bring about a common understanding of research integrity for the following reasons. In the first place, there was a growing concern about the increasing prevalence and, consequently, negative effects of misconduct. It cannot be denied that the effects of such misconduct are harmful indeed. It is damaging to *science*, because it may

[1]This paper is partly based on the report of Working Group 2 of the ESF (European Science Foundation) Member Forum on Research Integrity (ESF, 2010), and on an earlier publication of the author (Drenth, P. J. D. (2010). Research integrity: Protecting science, society and individuals. *European Review*, 18: 417–426.

create false leads for other scientists or the results may not be replicable, resulting in a continuation of the deception. It is also harmful to *individuals* and *society*: fraudulent research may result in the release and use of unsafe drugs, in the production of deficient products, inadequate instruments or erroneous procedures. Furthermore, if policy or legislation is based on the results of fraudulent research, harmful consequences are not inconceivable. But above all, damage is done through the subversion of the public's *trust in science*. The credibility of science would decline and trust in science as a dependable source of information and advice in respect of numerous decisions, so important for the welfare of mankind and society (environment, health, security, energy), would be subverted.

Academies of sciences and humanities in Europe have been involved in these discussions. In many countries the Academy is seen as an important actor in this arena. Academies are the scientific conscience of a society, and are considered an appropriate platform to consider issues of science and ethics, including research integrity and misconduct. Often they are expected to take the lead in defining and enforcing principles and standards for scientific integrity. In a policy briefing in 2002, the ESF envisaged an important task for the Academies of sciences and humanities in the formulation of national codes of good scientific practice and in the initiation of discussions on the most suitable approach to this problem.

Secondly, while international scientific collaboration increases sharply, another difficulty presents itself. Proper dealing with integrity and its obverse misconduct in an international projects is particularly difficult, as definitions, procedures and rules differ between the collaborating countries. Still, it is self-evident that common agreement on such standards, rules and procedures is a necessary precondition for the proper and responsible management of international projects. And it has to be admitted that many countries still lack a coherent and generally accepted policy and approach. Moreover, the present patchwork of codes and procedures is most inconvenient in international research projects.

It is self-evident that, in the same way that the national Academies are concerned with the national structures and policies with respect to research integrity, the European Federation of National Academies, *All European Academies (ALLEA)*, would focus its attention on the problems of integration and harmonisation of national codes, guidelines and best practices across Europe.

Integrity in science and scholarship became therefore a subject with high priority on the agenda of ALLEA's Standing Committee on Science and Ethics. During a number of its conferences and workshops and in a variety of publications, specific attention was paid to the problems of definition, identification, prevention and correction of various forms of infringements of the norms of scientific integrity. As a service to its members, in 2003, ALLEA published a "Memorandum on Scientific Integrity",[2] with suggestions how to identify and how to deal with scientific misconduct in their countries.

Like ALLEA, the *European Science Foundation* (ESF) has had a long-standing interest in integrity in science and the prevention and reduction of misconduct. In 2000, ESF published an important briefing on scientific practice in research and scholarship.[3] In 2008, ESF published a survey report[4] with an overview of various codes and guidelines to foster research integrity, procedures to deal with allegations of research misconduct, and key institutions to promote good research integrity in 30 European countries. Together with the US Office on Research Integrity (ORI), ESF organised the first World Conference on research integrity in Lisbon in 2007. This cooperation with ORI was continued with the preparation of the present second World Conference in Singapore.

In 2008, ALLEA and ESF decided to combine forces and to prepare (in collaboration with the UK Office of Research Integrity (UKRIO)) a project "European Coordinated Approaches to Research Integrity (ECARI)", to be financed by the European Commission. The objectives were, among others, to share information and experiences, to provide a vehicle for benchmarking best practices, to stimulate the development of appropriate structures, and to encourage the development of common approaches across Europe. Unfortunately, it took too much time and preparations to achieve a full and appropriate involvement of the European universities, and the funding could not be arranged. ALLEA and ESF then decided to pursue the ECARI objectives with their own resources.

2 WORKING GROUP "CODE OF CONDUCT" OF THE ESF MEMBER FORUM ON RESEARCH INTEGRITY

Within the framework of the ECARI project, ESF, together with the Spanish national research council CSIC, organised a workshop on research

[2]ALLEA (2003). *Memorandum on Scientific Integrity.* Amsterdam: ALLEA.

[3]ESF (2000). *Briefing on Good Scientific Practice in Research and Scholarship.* Strasbourg: ESF.

[4]ESF (2008). *Stewards of Integrity; Institutional Approaches to Promote and Safeguard Good Research Practice in Europe.* Strasbourg: ESF.

integrity titled "From principles to practice: How European research organisations implement research integrity guidelines". This workshop took place in Madrid on November 17–18, 2008. At this meeting a Member Forum on Research Integrity was established with the objectives to serve as a platform for the exchange of information on attempts and initiatives to ensure research integrity and to prevent misconduct, and to encourage organisations which do not yet have appropriate structures to initiate debates in their respective communities on adequate models. Four working groups were charged with the task of each addressing one particular aspect of the problem area.

The four working groups and their commissions were:[5]

- WG 1 "Raising awareness and sharing information" (chair: Sonia Ftacnikova (SL)). The task of this working group was to develop and implement activities to continue raising awareness and sharing information on good practices to promote and safeguard research integrity.
- WG 2 "Code of Conduct" (chair: Pieter Drenth (NL)). This working group was to develop a Code of Conduct which defines core values to be pursued and norms to be complied with in responsible research, and which could then be used as a template for national or institutional codes of conduct in Europe.
- WG 3 "Setting up national structures" (chair: Maura Hiney (IE)). This working group had to develop a checklist for setting up national and institutional structures to promote good research practices and deal with research misconduct.
- WG 4 "Research on scientific integrity" (chair: Livia Puljak (HR)). This working group had to develop and promote research programmes to take stocks of what is already known and better understand research misconduct (occurrences, contributing factors, effectiveness of various measures, etc.).

The four working groups worked in collaboration and integrated their insights and conclusions in a comprehensive strategy for promoting and safeguarding integrity in scientific and scholarly research and practice, both nationally and in the wider European context.[6] Their final report "Fostering Research Integrity in Europe" appeared in the fall of 2010. (www.esf.org).

[5]For information on the other members of the Working Group 2, see Appendix I.
[6]The Member Forum on RI is thankful to Laura Marin (ESF) for her coordinating activities.

The Code of Conduct proposed by Working Group 2 emerged from a series of discussions both within WG2 and ALLEA on the basis of a preliminary discussion paper.[7] Evaluation and feedback were given by ALLEA's Standing Committee on Science and Ethics, and by representatives of ALLEA's Member Academies at a special meeting in Berne (June 29–30, 2009). Each subsequent version was discussed and commented on by WG2. In this manner, the final proposal of the ESF Member Forum has also met with the general approval of the European national Academies, associated with ALLEA.

3 THE CODE

For the full text of the European Code of Conduct for Research Integrity, the reader is referred to the ESF/ALLEA publication "Fostering Research Integrity in Europe" (Strasbourg, 2011). Here we present a short recapitulation of the main points of this Code.

Principles of integrity. Researchers, public and private research organisations, universities and funding organisations must observe and promote the principles of integrity in scientific and scholarly research.

These require honesty in presenting goals and intentions, in reporting methods and procedures, and in conveying interpretations. Research must be reliable and its communication fair and full. Objectivity requires facts capable of proof, and transparency in the handling of data. Researchers should be independent and impartial, and communication with other researchers and with the public should be open and honest. All researchers have a duty of care for the humans, animals, the environment or the objects that they study. They must show fairness in providing references and giving credit for the work of others, and must show responsibility for future generations in their supervision of young scientists and scholars.

Misconduct. Research misconduct is harmful for knowledge. It could mislead other researchers, it may threaten individuals, or society — for instance, if it becomes the basis for unsafe drugs or unwise legislation — and by subverting the public's trust, it could lead to a disregard for or undesirable restrictions being imposed on research.

[7]Drenth, P. J. D. (2009). *Scientific Integrity,* discussion paper Amsterdam: ALLEA.

Research misconduct can appear in many guises:

- *Fabrication* involves making up results and recording them as if they were real.
- *Falsification* involves manipulating research processes or changing or omitting data.
- *Plagiarism* is the appropriation of other people's material without giving proper credit.
- Other forms of misconduct include *failure to meet clear ethical and legal requirements*, such as misrepresentation of interests, breach of confidentiality, lack of informed consent, and abuse of research subjects or materials. Misconduct also includes *improper dealing* with infringements, such as attempts to cover up misconduct, and reprisals on whistleblowers.
- *Minor misdemeanours* may not lead to formal investigations, but are just as damaging given their probable frequency, and should be corrected by teachers and mentors.

The response must be proportionate to the seriousness of the misconduct: as a rule it must be demonstrated that the misconduct was committed intentionally, knowingly or recklessly. Proof must be based on the preponderance of evidence. Research misconduct should not include honest errors or differences of opinion. Misbehaviour such as the intimidation of students, misuse of funds and other behaviour that is already subject to universal legal and social penalties is unacceptable as well, but is not "research misconduct" since it does not affect the integrity of the research record itself.

Good Practices. There are other failures to adhere to good practices — incorrect procedures, faulty data management and so on — that may affect public trust in science. These should be taken seriously by the research community as well. Accordingly, data practices should preserve original data and make it accessible to colleagues. Deviations from research procedures include insufficient care for human subjects, animals or cultural objects; violation of protocols; failure to obtain informed consent; breach of confidentiality etc. It is unacceptable to claim or grant undeserved authorship or deny deserved authorship. Other publication-related lapses could include repeated publication, salami-slicing or insufficient acknowledgement of contributors or sponsors. Reviewers and editors too should maintain their independence, declare any conflicts of interest, and be wary of personal bias and rivalry. Unjustified claims of authorship and ghost authorship are forms of falsification. An editor or reviewer who purloins ideas commits plagiarism.

It is ethically unacceptable to cause pain or stress to those who take part in research, or to expose them to hazards without informed consent.

While principles of integrity and the violation thereof, have a *universal* character, some rules for good practice may be subject to cultural differences, and should be part of a set of national or institutional guidelines. These cannot easily be incorporated into a universal code of conduct.

International collaboration. In international collaboration partners should agree to conduct their research according to the same standards of research integrity, and to bring any suspected deviation from these standards, in particular alleged research misconduct, to the immediate attention of the project leader(s) (and of the senior responsible officer in the university or institute (employer)). This allows for investigations according to the policies and procedures of the partner with the primary responsibility, while respecting the laws and sovereignty of the States of all participating parties. In large-scale, funded international projects, the promotion of good practice and the handling of possible cases of misconduct, as recommended by the co-ordinating committee of the OECD Global Science Forum, should be followed. The boiler plate text, recommended by this committee, should be embodied in the formal documents that establish the collaborative project.

4 SOME ANNOTATIONS

With respect to this European Code, I would like to make the following comments:

- This Code of Conduct is not a body of law. It is not intended to have a legal character, but rather to be a canon for self-regulation. It is a basic responsibility of the scientific community to formulate the principles and virtues of scientific and scholarly research, to define its criteria for proper research behaviour, and to set its own house in order when scientific integrity is threatened.
- The Code applies to all fields of science and scholarship. Although natural sciences, life sciences, social sciences and humanities may differ in methods and traditions, they all have a fundamental characteristic in common: they depend on argument and evidence, i.e. observations of nature, or of humans and their actions and products.
- Science and scholarship are embedded in a wider socio-ethical context. Scientists and scholars have to be aware of their specific responsibility towards society and the welfare of mankind. For example, they bear

responsibility for the choice of subjects to be investigated and its consequences, and also for the practical applications and uses of their research results. In this Code, however, we confine ourselves to standards of integrity while *conducting* research, and do not consider this wider socio-ethical responsibility.

- The objective of the European Code is to stimulate and develop the emergence of institutional settings that strengthen research integrity, and to set standards across Europe that can, eventually, be held valid and implemented worldwide.
- The European Code could:
  - function as a benchmark for proper behaviour of researchers in international collaborative research and for institutions that want to develop their own code;
  - be a basis for the development of national regulations where none exist;
  - complement existing codes of ethics; and/or
  - in some cases enhance or supercede codes already in operation.
- This Code represents agreement at a given point in time. Changing national or institutional frameworks, or scientific or technological developments, may make some adjustments necessary.
- The confinement to a European Code of Conduct does not imply that the principles and guidelines are to remain restricted to the European community. It is our hope that part of our work could be useful for other parts of the world as well.

5 APPENDIX I. MEMBERS OF THE WORKING GROUP

Member	Organisation	Country
Pieter Drenth (chair)	All European Academies (ALLEA)	NL
Tommy Dahlen	Swedish Council of work life and social research	SE
Glyn Davies	Economic and Social Research Council (ESRC)	UK
Kirsten Hüttemann	Deutsche Forschungsgemeinschaft (DFG)	GE
Pavel Kratochvil	Academy of Sciences of the Czech Republic	CZ
Michelle Hadchouel	Institute Nationale de la Santé et de la recherche Medicale (INSERM)	FR
Pere Puigdomènech	Consejo Superior de Investigaciones Cientificas (CSIC)	SP

CHAPTER 23

A REPORT FROM THE EUROPEAN FORUM FOR GOOD CLINICAL PRACTICE

Frank Wells

The European Forum for Good Clinical Practice (EFGCP) is a non-profit organisation established by and for individuals with a professional involvement in the conduct of biomedical research, and its purpose is to promote good clinical practice and encourage the practice of common, high-quality standards in all stages of biomedical research throughout Europe. Its 2009 Annual Conference was entitled "Research Integrity: A European Perspective". After two days of discussion and debate, a number of conclusions were reached, which were taken forward by a small international multidisciplinary group of interested persons ("The Group"). A full report on their deliberations is published on the EFGCP website (www.efgcp.eu) of which this is a summary.

Definitions of "fraud" and, particularly, "misconduct", were needed with clear demarcations between them, across Europe and the rest of the world. The Group concluded, after much discussion, that language differences between various European countries, especially in translation, were such that it would be better to derive just one comprehensive and unambiguous definition of research misconduct, indicating within that definition where fraud applied. Essentially, when the misconduct was clearly *intended* to obtain advantage or to deceive, that was defined to be fraud.

The case had been established in Denmark, the Nordic Countries and in the USA (with the ORI) for a National Body on Research Integrity. The case now had to be made for establishing such bodies in other countries.

The Group therefore recommended that:

- An independent national panel should be established — where one does not already exist — with public representation, to provide advice and assistance on request.
- Such a panel should develop and promote models of good practice for local implementation; provide assistance with the investigation of alleged research misconduct (including a "rapid response" facility); collect, collate and publish information on instances of research misconduct; and publish an Annual Report in both English and the national language on the cases that have been investigated.
- All stakeholders involved in clinical research should ensure that the public record is corrected if found to be incorrect.

The Group felt strongly that training in the detection and investigation of suspected research misconduct should be offered, and be expected to be taken up, by all stakeholders involved in clinical research, including research ethics committees. Even more important is training in the principles of research integrity and the *prevention* of fraud and misconduct. This cannot be over-emphasised, and yet the research conducted by The Group had failed to reveal any consistent training programmes on the principles of research integrity virtually anywhere in Europe.

Support was needed for research into research misconduct, as the true prevalence of the various grades of research misconduct was not known, nor did any guidelines exist on what was acceptable and what was unacceptable.

The Group felt that it was essential to establish internationally agreed guidelines for whistleblowers, given that the fate of whistleblowers is generally not an easy one to accept. A draft version of such guidelines appears on the EFGCP website, providing information on the following topics:

- What is whistleblowing?
- Raising concerns: the principles
- Types of concern
- Concerns about research
- What are the contractual entitlements?
- What are the professional obligations?
- Who do I approach in order to raise a concern?
- Raising a concern
- Will there be personal consequences if concerns are raised?
- Where else can guidance be found?

The Group felt strongly that similar guidance should be available for any organisation involved in the conduct of clinical trials, but appreciated that the guidance would need to be adapted so as to be suitable for each different organisation. Nevertheless, it felt that the publication of guidance material for a pharmaceutical company could be used as a template for other organisations, and such a document therefore appears on the EFGCP website. The document includes the information on such guidelines:

- Objective
- Scope
- Applications
- Policy
- Procedure
- Training

It is always a quality control failure if fraud or misconduct is first identified during an audit. The operational checks by those responsible for the work, such as auditors and data managers, should identify such serious issues as part of their routine assessment.

It is often an automatic response to suspected fraud for the client group to request an audit. This has the advantage of showing that the sponsor takes these issues very seriously, but it is not a regulatory requirement that an audit must be conducted in every case. When the request comes in, it is better to take some time to understand exactly what has been noticed that has caused the concern. When the audit team are fully informed, they can then consider the question, "What can the auditors do in this situation that is not possible for the quality control staff?"

The role of the statistician in confirming or denying a suspicion that data have been fabricated or falsified is under-appreciated. Good guidelines, already published in all four editions of *Fraud and Misconduct in Biomedical Research* (Wells and Farthing (eds.)), need to be promulgated. The complete chapter, written by Professor Stephen Evans, is reproduced within the report on the EFGCP website.

The Group deliberated over whether there should be any reference to the establishment of Rapid Response Teams of experienced independent forensic investigators, such as had been in existence in the United Kingdom for a number of years. It decided that it would be appropriate to make a case for the establishment of such a team or teams, and an article which sets out the experience and validity of one such team is shortly to appear in the *Journal of the Royal Society of Medicine.* A resume of this article can also be found on the EFGCP website.

In the context of the conduct of clinical trials on medicinal products, national competent authorities (NCAs) must be involved, and their role in regulating the development and marketing of medicinal products covers a wide range of activities and regulated parties. Misconduct can be carried out within any of the activities and by any of the parties involved. The NCAs have a key role to play in the prevention of misconduct, in the investigation of potential cases when they occur, and in taking action or initiating action by other authorities where there is evidence of a potential misconduct. However, their current commitment throughout the world is variable, and The Group therefore felt there was a need for harmonisation between NCAs, particularly in the context of misconduct within a multinational trial. In order for an effective regulatory framework to function, a number of elements need to be in place. These are set out in the complete report of The Group which is on the EFGCP website: www.efgcp.eu.

CHAPTER 24

LESSONS FROM 17 YEARS WITH NATIONAL GUIDELINES FOR RESEARCH ETHICS IN NORWAY

Ragnvald Kalleberg

The World Conference organisers have asked us "... to discuss the goals, content, and use of codes of conduct developed by research institutions and professional or academic societies". We have been asked to address four questions: (1) What is the purpose of codes at the institutional level? (2) What topics should be covered? (3) Should codes play a regulatory role? (4) What problems arise when adopting institutional codes? In order to be grounded in concrete experiences, this paper is focused on some lessons from a national code in Norway, namely the *Guidelines for Research Ethics in the Social Sciences, Law and the Humanities* (see internet address). The guidelines were first adopted in 1993, and have been revised two times, in 1999 and 2006.

The Norwegian Guidelines were developed by researchers for researchers. The 47 guidelines were formulated and published by one of the three national committees for research ethics. The three cover the main scholarly and scientific fields: natural science and technology, medical research, and social sciences and humanities. Each committee has 12 members comprising 10 researchers and two lay members. They are advisory bodies — final responsibility rests with the research institutions. The Guidelines were formulated by experienced researchers from the relevant disciplines. The guidelines have the form of prescriptions (for instance about free and informed consent), proscriptions (for instance against fabrication of data) and recommendations (for instance about training in research ethics). These norms have two sources: 1) they are explications of implicit norms followed or condemned in adequately functioning research communities, and 2) they are articulated on the basis of problematic cases, for instance examples of scientific misconduct. Each revision of the Guidelines has been accompanied by comprehensive national hearings, with extensive and detailed critical and constructive responses from scientific institutions.

The Guidelines reflect a broad conception of research ethics (research integrity). The guidelines can be organised into four groups: (1) Norms for freedom of research, good research practice associated with scientific quality, and norms regulating relations between researchers; (2) Norms regulating relationships to the subjects that are being studied (individuals, groups, institutions); (3) Norms regulating relations to those financing or ordering the research (contract research) or to users such as employers, employees, customers and clients; (4) Norms, values and guidelines relevant for the communication of scientific knowledge to society outside of the specialised research fields (researchers in other specialised fields included). The relationships to citizens include: (a) science as a cultural value in society; and (b) science as a public resource to promote open and enlightened public discourse in deliberative democracy (cf Kalleberg, 2010).

The Guidelines apply to both individuals and institutions. The purpose of the guidelines is to help research communities promote good practice and prevent misconduct. It is understood that research ethics is basically an institutional phenomenon, comprising individuals as well as interaction in groups and institutions. Guideline 4, for example, states, "Institutions and individual researchers shall develop and maintain good research practices. Institutions are to have procedures to handle violations of ethical research norms" (my revised translation). Another example is to be found in the last section, on the communication of science to the public at large. Guideline 42 sets out requirements for individual researchers and research groups as well as requirements to "institutions to pave the way for multifaceted, comprehensive science communication".

Acceptance of the Guidelines. The national guidelines are widely known among Norwegian researchers. They have high legitimacy in research institutions. I think this, to a large degree, is based on the fact that they are articulated by experienced research practitioners and addressed to other research practitioners and students. Research ethics is not understood as something that is forced on research communities from the outside. It is an internal connection between epistemology and ethics. The guidelines are widely understood as clarifications or explications of actual good practices or resulting as general insights from extensive discussions — also in the larger public — for instance about plagiarism. Most of the guidelines are *prima facie* norms, having to be weighed in relation to other norms and considerations.

Use of the Guidelines. The guidelines play a regulatory role in discussions in scientific communities when research ethical challenges are discussed, as well as when groups and institutions ask the national committee for advice. The guidelines are extensively used in education at the graduate level.

References

1. Kalleberg, R. (2010). "The ethos of science and the ethos of democracy". In *Robert K. Merton: Social Theory and Sociology of Science*, Calhoun, C. (ed.), pp. 182–213. New York: Columbia University Press.
2. *Guidelines for Research Ethics in the Social Sciences, Law and the Humanities.* (First version 1993. Third revised edition 2006).
3. *Guidelines for Research Ethics in Science and Technology.* (First version 2007).
4. National Committees for Research Ethics in Norway. (www.etikkom.no) www.etikkom.no/en/In-English/Publications/

CHAPTER 25

SOCIETY FOR SCIENTIFIC VALUES: A MOVEMENT TO PROMOTE ETHICS IN THE CONDUCT OF SCIENCE[1]

Ashima Anand

Before the malaise of publish or perish, which could have spawned an epidemic of scientific misconduct, hit the younger generation of scientists in India, cases of blatant misdemeanours involving renowned scientists had already started coming to light. One of the challenges in dealing with this was the lack of interest from the scientific bureaucracy. It was either oblivious of the long-term implications of such episodes, or it just preferred to turn a blind eye to them. It also appeared that most of the bureaucracy was on the same side of the power equation as the perpetrators.

To deal with this problem, an eminent group of concerned scientists and engineers came together in the early 1980s and, after much deliberation with colleagues country wide, they set up a society in 1986, which was registered and named the "Society for Scientific Values" (SSV). Its first President was Professor A. S. Paintal, FRS. The mandate of this society was to promote integrity, objectivity and ethical values in the conduct of scientific activity. Its chief aim: to evolve a healthy scientific environment, which would be free from prejudices, bureaucratic formalities, dishonesty, propagation of unsubstantiated research claims, suppression of dissent, showmanship, sycophancy and political manipulation, all of which engender misconduct in scientific activity.

[1]I am grateful to Professor KL Chopra, President SSV for his invaluable inputs. (choprakl@yahoo.com).

1 HOW SSV WORKS

SSV's membership, which now numbers over 400 scientists, is open only to those who have a clean record and who are actually concerned and serious in promoting the society's ideals in the organisation in which they work. New members are nominated by the existing membership on the basis of their reputation and standing, not only in their respective fields but also of their known concerns and efforts in this direction. To elaborate, they must meet the following criteria:

1. He/she should have allowed his name to appear as an author only in those publications in which he/she was actively involved, e.g. in data collection, theoretical formulation, design and construction of apparatus, field trips, mathematical derivation and calculations, statistical analysis and interpretation of results, as distinct from administrative support and providing funds or facilities.
2. He/she should never have plagiarised or made false claims or indulged in or supported and encouraged any kind of unethical activity in science.
3. He/she should wholeheartedly support the decisions and actions to be taken collectively by the Society after such decisions and actions have been approved.

Finally, the Society makes it quite clear that members should agree to withdraw from the Society if they cease to adhere to guidelines 1, 2 and 3 above.

2 INVOLVEMENT OF MEMBERSHIP IN INQUIRING CASES OF MISCONDUCT

Since there are no existing codes or a redress system in place, the Society is continually being contacted about cases of alleged scientific misconduct that are taking place in universities and research institutions across the country. In fact, it has taken on the role of being the "court of last resort" in such matters. It conducts an inquiry in a three-step manner. For example, in cases of alleged plagiarism, after hearing from a whistleblower, it notifies the accused and the victim as well as the head of the institution where the misdemeanour was supposed to have taken place. It then awaits the details of their stand on it. In the meantime, a committee is set up comprising of SSV members who are experts in the subject concerned. All available documents in the matter are thoroughly scrutinised by this committee and lead to a report. This is conveyed to the three parties with advice to the head of the institution to take appropriate action. One of the SSV's recent and major achievements has been the removal of three heads of institutions charged with plagiarism. It took over two years of constant pressure on the concerned authorities in each case to achieve this.

3 ROLE OF SSV WEBSITE AND NEWSLETTER

Once the report has been finalised, it is posted on the Society's website (scientificvalues.org). Thus its website has become an active part of the effort to promote integrity in research in India. In addition to this, twice a year SSV publishes a newsletter titled "News & Views", which is a detailed update of its activities. This content is also available on its website, and is sent around to its members and also to the libraries of institutes, academies and universities.

In the past, it was not always possible to take action against all those found guilty of scientific misconduct by the Society's initiative. But as this process has started to become better known for its well-intentioned initiative and judiciousness, several heads of institutions have begun to take note of it and have actually been referring back to the President of the Society for advice about the possible disciplinary action to be taken against the culprits under their jurisdiction.

4 SENSITISING SCIENTISTS

In the absence of the existence of codes, the Society has also undertaken to sensitise the young research scientists by organising national symposiums on various issues that constitute this problem. The first one was held in 2003 and dealt with Scientific Values and Excellence in Science. Others that were held over the years dealt with:

- Accountability in Scientific Research
- Scientific Misconduct and Disciplinary Action
- Ethics in Administration of Science
- Accountability in Scientific Institutions and Discussion on Corruption and Whistleblowers Protection Act

Keeping in view that "ethics" cannot be taught in a formal way but must be cultured and nurtured through experience, analysis, introspection, and a sense of responsibility, SSV encourages its members to continually talk about scientific integrity and elaborate on codes of conduct when they visit institutes or universities for other business such as symposiums or scientific meetings. Over the last three years, 20 such institutional seminars have been held in the country with an excellent interactive response from students and faculty.

A few academic institutions have already set up Centres for Value Education, which offer courses designed for motivating interested students in ethics. The University Grants Commission (UGC) has, under persuasion from the President of the Society, finally agreed to advise all universities to conduct a two-credit course on ethical values in science and technology. A committee of the SSV is working on preparing suitable modules of all case studies of plagiarism that it has resolved, as material for such a course.

Although it lacks administrative or legal power, the Society for Scientific Values continues to achieve a lot — its roles as a co-whistleblower and moral watchdog are now well established in India and across the world. SSV is especially known to the editors of peer-reviewed journals.

CHAPTER 26

HOW MANY CODES OF CONDUCT DO WE NEED? THE CHINESE EXPERIENCE

Ping Sun

In China, many codes of research conduct were developed by government departments, universities, research institutions, the national academies and learned societies. These came in the form of guidelines, self-disciplinary principles and norms for special activities. Their target audiences include all science workers, faculty and students, researchers, academicians, journal editors, and others.

A typical code of conduct consists of several parts: general principles, norms of research ethics, definitions of research misconduct, and the handling of research misconduct. Besides numerous institutional codes of conduct, there are some "national" ones, including *Some Opinions on the Code of Conduct for the Science Workers* issued by the Ministry of Science and Technology, Ministry of Education, Chinese Academy of Sciences, Chinese Academy of Engineering, and China Association for Science and Technology (CAST) in 1999, and *The Norms of Scientific Ethics for the Science Workers (Trial)* drafted and adopted by the CAST in 2007.

As these codes or guidelines are developed with similar goals, universal principles and contents, several important questions arise, including: Is it necessary for every institution to develop its own code of conduct? Which codes of conduct are better than others, if there are better ones? If there are inconsistencies between different codes, how are people supposed to act? And most importantly, how many codes of conduct do we really need? With an informal review of some typical codes of conduct and drafting processes, the author suggests that a small number of professional or disciplinary codes of conduct is better than a large number of institutional ones.

1 IT IS EFFICIENT AND EFFECTIVE TO DEVELOP PROFESSIONAL OR DISCIPLINARY CODES OF CONDUCT

As faculty, students, and researchers work in different disciplines and fields, institutional codes of conduct tend to be comprehensive to suit all its members, while some content in institutional codes will not be applicable to all members of the community. There are many institutions in China that are still developing or revising their codes of conduct, sometimes adopting existing templates with minor alterations, and at other times carrying out relevant research and putting forward new versions. It is not easy for untrained individuals to develop codes, and revising work can also be demanding. It is therefore desirable for appropriate organisations (learned societies, professional associations) to develop codes of conduct in major professions and research disciplines, to reduce the burden on individual institutions by responding to suggestions from the stakeholders and eliminating all possible inconsistencies.

2 IT IS IMPORTANT TO MAKE SURE THAT THE CODES OF CONDUCT ARE AUTHENTIC

There are imperfections in many codes of conduct in China. For example, though both encouraged and discouraged behaviours are usually listed in a code, some behaviours are not well defined, or are defined differently in various codes. And as it is difficult to determine what should be included in a code and what should be left out. The mixture of principles and details in some codes leads to a sense of imbalance. In order to avoid confusion and inconsistencies, it is important to make sure that the codes of conduct are authentic, in accordance with relevant laws and regulations, and widely accepted by the scientific community. Useful codes should reflect the most important, agreed-upon principles and essential requirements for target audiences, and should be supplemented by detailed research norms in different disciplines and fields.

3 THE MAJOR CHALLENGES FOR DEVELOPING DESIRABLE CODES OF CONDUCT

The first challenge is "authority". When developing a code of conduct, the drafting body has to consider: Is our organisation entitled or authorised to do so? Is the procedure of drafting, reviewing and adopting the code justified? Some organisations cannot answer these questions confidently, so they choose the term "guidelines" instead of code of conduct, and aim to limit its jurisdiction.

The second challenge is the setting of standards. There are many open questions without definite answers. For example, what should be the appropriate policy for data management? What is the definition of "conflict of interest"? What responsibilities should be assumed by universities or research institutions? Efforts to reach a consensus on these issues has proven to be difficult for relevant organisations and researchers.

The third challenge is implementation. Though a code of conduct is usually well-publicised at the time of issuance, it is often not clear that the target audience are aware of its existence, or whether their behaviour is affected by it. Though policy and procedures for handling research misconduct are included in some institutional codes in China, but due to lack of operational details, some codes are mainly of encouragement and discouragement nature.

4 STRATEGIES FOR DEVELOPING CODES OF CONDUCT

There are many options for developing codes of research conduct. Considering the imperfection of the current codes of conduct, an ideal code should be concise, well-drafted, and preferably one or two pages long, to enhance the understanding of its audience and reduce their burden for remembering all the content. An example of such code is the *Hippocratic Oath* for medicine, which highlighted the most important principles and norms at that time. It is better to develop an international code of conduct, as there are seldom any differences in essential research norms around the world, thus the *Singapore Statement on Research Integrity* is a good initiative.

The following strategies can be taken to develop a code of conduct. The appropriate organisations should take the lead (1) to develop codes of conduct and invite the participation of relevant organisations if necessary; (2) to assess the utility and effectiveness of similar existing codes to gain experience; and (3) to develop and promulgate detailed norms and regulations at the same time, for science workers to refer to when they need to consult the sure about the standards and other technical details.

Besides the code of conduct, certain template policies and procedures for handling research misconduct should also be developed, with the efforts led by government agencies, to provide guidance for relevant institutions.

CHAPTER 27

THE PAST, PRESENT AND FUTURE OF ONE UNIVERSITY'S CODE OF RESEARCH ETHICS IN NEW ZEALAND

John O'Neill and Sylvia Rumball

Massey University is one of eight universities in New Zealand. In 2009, it had approximately 35,000 students, nineteen percent of whom were postgraduates. A third of its operating income was from research. Over three quarters of the academic staff were "research active".

University research activity and oversight have increased hugely since 1980 when the Vice-Chancellor's Office had only two assistants and a secretary, and the university, nine research centres. By 1990, there were committees for animal, genetic and human research ethics.[1] The Vice-Chancellor's Office had six assistants, including one in change of research. There were 12 research centres. In 2010, the Office of the Assistant Vice-Chancellor for Research alone employed 75 FTE staff, with units for research management, ethics and a graduate research school. There were 51 research centres.

The Code of Human Research Ethics was developed in consultation with staff between 1988 and 1990. It was short and relatively non-prescriptive, with an emphasis on education. Researchers were trusted to act ethically within a culture of collegial disciplinary groupings and academic service. The external funding and regulatory environment was benevolent.

By 2010, the ratio of general to academic staff had increased significantly, due to a mixture of compliance and entrepreneurial requirements.

[1]The Massey University Codes covering the ethics of animal and genetic technology research are largely prescribed by the procedural requirements of specific acts of parliament and regulations and, therefore, are not considered in this paper.

Three codes of research ethics are now referenced within the University's research conduct policy[2] and form part of the individual's employment agreement.

The focus of staff energy is on personal research "outputs", while service contributions have diminished. Capped public funding has caused a growth in "commercial science". The main researcher affiliation may no longer be discipline or workgroup, but, increasingly, external "reputation".

By 2030, the changing balance between public and private revenue is predicted to create ethical dilemmas of (a) funding and (b) "intellectual property". The 1989 Education Act required universities to remain "disinterested".[3] In order to fulfil this role as a public good, they were guaranteed institutional autonomy and academic freedom. By 2030, the challenge will be to find private funding to maintain the integrity of public research and teaching.

This will require a radical shift in conception of what university codes that apply to individual researcher conduct should do. In such circumstances, the focus of codified conduct must shift from the individual to the institution as a whole, and from individual researcher ethics to the integrity of corporate governance.

RECOMMENDATIONS

Recognise that "research ethics" and "research integrity" are culturally-valued forms of academic decision-making. However, as teaching and research demands grow, service contributions to academic decision-making decrease. This potentially distorts academic trust and identity. Universities will therefore need to decide whether future codes should more clearly specify the individual's obligations and duties.

Accept that universities, as public good institutions, must find alternative sources of funding to supplement the diminishing state subsidy. Most universities have yet to articulate the criteria against which they will accept or invest in research funding. To strike a balance between commercial and open science commitments, they will need to promote corporate "integrity" of research decision-making at the highest governance level of the institution.

---

[2]Massey University (2007, January). *Code of Responsible Research Conduct and Procedures for Dealing with Misconduct in Research.* Palmerston North, NZ: Author.

[3]NZ Parliament. *Education Act (1989)*, Sections 161 and 162.

CHAPTER 28

THE AUSTRALIAN CODE FOR THE RESPONSIBLE CONDUCT OF RESEARCH — CHALLENGES AND RESPONSES

Timothy Dyke

Australia's framework for the conduct of research considers research governance and conduct, human research and animal research. Some areas such as gene technology, therapeutic goods and human embryo research are covered by legislation. The framework is based on national standards and guidelines. Responsible conduct of research is largely an institutional responsibility. Requirements are expected to be followed, and compliance is monitored through funding body agreements.

There are three Australian national standards co-authored or endorsed by the National Health and Medical Research Council, Australian Research Council and Universities Australia. The three standards are:

- The Australian Code for the Responsible Conduct of Research (2007);
- The National Statement of Ethical Conduct in Human Research (2007); and
- The Australian Code of Practice for the Care and Use of Animals in Scientific Purposes (2004).

1 AUSTRALIAN CODE FOR THE RESPONSIBLE CONDUCT OF RESEARCH 2007

Part A of the Code covers

- responsibilities of institutions including promoting the responsible conduct of research, establishing good governance and management

practices, training staff, promotion of mentoring and ensuring a safe research environment
- responsibilities of researchers — researchers should maintain high standards of responsible research, report research responsibly, respect human research participants, consider involvement in or effects of research on Indigenous Australians, respect animals used in research, respect the environment, report research misconduct
- consumer and community participation in research
- data management
- supervision
- publication
- authorship
- peer review
- conflicts of interest
- collaborative research

Part B of the Code outlines a framework for the handling allegations of research misconduct.

2 POSITIVE ASPECTS

Joint authorship of the Code has demonstrated a common understand of good research practices. Institutional responsibilities and researcher responsibilities are clearly outlined. The need for an external and independent process for considering serious research misconduct is a strong aspect of the Code.

3 CHALLENGES

Processes outlined in the Code for handling allegations of serious research misconduct and processes described in employment agreements with employees are sometimes at variance. Handling of allegations involving non-employees (students, honoraries, visiting faculty), previous employees (retired, resigned) has presented challenges for some institutions. Other challenges that institutions have faced in implementing the Code have included

- the need for better governance arrangements for collaborative research;
- responsibilities to raise awareness of the Code and to investigate allegations of research misconduct; and
- avenues for complaint or appeal

4 OTHER AUSTRALIAN ADVANCES IN RESEARCH GOVERNANCE

The Australian Research Integrity Committee will be established by the government to consider cases in which persons have concerns over the processes used by an institution in the handling of allegations of research misconduct. Further information is available at www.nhmrc.gov.au.

SECTION V

INSTITUTIONAL AND NATIONAL APPROACHES TO FOSTERING RESPONSIBLE RESEARCH

INTRODUCTION

As noted in the Workshop Report included in Section 6, the need for formal Responsible Conduct of Research (RCR) education has been recognised in the USA since the 1990s. However, global recognition of the need for training has developed more slowly. Countries that have in one way or another addressed the issue of research integrity frequently mention the importance of RCR education, but to date no countries have followed the USA lead and adopted formal requirements. Consequently, formal RCR education is not available to most research students or new researchers outside the USA. Even when it is offered, uneven and inconsistent RCR education poses a significant challenge, particularly for international research collaboration. These shortcomings open opportunities for working together and developing innovative ways to make researchers aware of regulations and other responsibilities.

Kalichman opens the discussion of institutional and national ways to foster responsible research with a thoughtful and challenging analysis of the wide variation in goals, approaches and content for RCR teaching. While it might seem that finding common ground in this area should not be a problem, based on his own and other studies, he reports that the terms that instructors use to identify ideals and problem behaviours, the topics they cover, their audiences and more can share little in common. He therefore ends his analysis with a series of questions that need to be addressed in order to move forward.

In his paper, Langlais shifts the focus of promoting integrity to institutional culture. Fostering integrity requires buy-in from the community, to dispel myths about behaviour and the amount of training that is already available. He began his efforts as a research administrator by setting up a committee made up of respected faculty and campus leaders and by studying the climate at his own institution. One important discovery was that faculty and students differed on how much training was already provided. Faculty felt it was adequate, students that it was not. This provided the impetus for establishing campus-wide RCR training.

Sponholz describes how a curriculum for "Good Scientific Practice" in science and medicine was developed on behalf of and in cooperation with the ombudsman of the Deutsche Forschungsgemeinschaft (DFG). RCR training is currently optional in Germany. The curriculum is broken down into four parts: background, principles, practical training and assessment, delivered at two levels: beginning and mid-level. It provides both content outlines and suggested resources for teaching RCR. Two obstacles, a lack of teachers and a lack of commitment, now hamper broad implementation, leading some to suggest that such training should be made compulsory.

Kalleberg prefaces his discussion of developments in Norway with a view that ethics should be seen from an internal research perspective not imposed from the outside, and as more institutional than individual. Norway's interest in RCR training increased significantly after the Sudbϕ case and led to significant efforts to improve both environment and training. However, Kalleberg suggests that more effort is needed. Less than 5% of researchers he surveyed had learned about research integrity from their mentor.

Vasconcelos raises special challenges for RCR training in Brazil and other similar countries, with language and cultural difference. Although a growing research power, many of Brazil's researchers do not easily read and write English, putting them at a disadvantage when it comes to publishing their work in major scientific journals. Some Brazilian researchers also have different attitudes to best practices, such as their attitudes toward copying text from others in some parts of publications. For RCR training to be more effective in Brazil, Vasconcelos points to the need for more RCR instructors, more room for RCR in the curriculum and clearer policies.

Examples of the different ways individuals have responded to the challenge of fostering responsible research are described in the next section, along with the report from an international workshop on this topic held the day after the World Conference.

CHAPTER 29

WHY, WHAT, AND HOW WE SHOULD BE TEACHING ABOUT RESEARCH INTEGRITY[1]

Michael Kalichman

The purpose of this background paper is to serve as a framework to discuss rather than to definitively answer the questions of why, what, and how for the teaching of research integrity. As will hopefully be made clear, we have considerable experience with these issues in the USA, but we certainly have not yet fully met the challenge of how best to promote research integrity. Given the increasingly multi-national context of research, it is clear that we nominally should aspire to seeing our role as *global* citizens committed to carrying out research with the highest standards of integrity. Before further examining this goal, it is important to first be clear about a few definitions.

1 Definitions

Many different terminologies have been used for what often amounts to the same thing seen from different vantage points: research ethics, research integrity, scientific ethics, scientific integrity, responsible conduct of research (RCR), good scientific practice, and others.

In the USA, there is no one accepted name for this subject and, often for very good reasons, scholars and policymakers find important distinctions

[1]The preparation of this overview was supported by Grant Number NR009962 from the National Institute for Nursing Research. Drs. Elizabeth Heitman, Brian Martinson, Dena Plemmons, and Daniel Vasgird are thanked for many helpful discussions and suggestions. The content is solely the responsibility of the author and does not necessarily represent their views or the official views of the National Institute for Nursing Research or the National Institutes of Health (NIH).

among these terms. For some, "research" might include a wide range of scholarly pursuits (e.g. philosophers and historians conduct research), while "science" or "scientific" might be limited to those whose work is based more on observation and experiment (e.g. botanists and physicists). Because of a prominent historical focus on the ethics of research with human subjects, some reserve the term "research ethics" for only that kind of research that is for the purpose of studying humans (Dresser, 2001; Iltis, 2006; Israel and Hay, 2006). And the terminology of "responsible conduct of research" (RCR) is variably considered to be synonymous with the NIH Training Grant requirement for RCR education (NIH, 1989), for the topics covered by the PHS policy for RCR training proposed in December 2000 (PHS, 2000) and suspended in February 2001 (DHHS, 2001), the recent announcement of an NSF policy requiring RCR education (NSF, 2010), or the clarifications of the NIH RCR requirement (NIH, 2009). Despite their use of the same terminology (i.e. "RCR"), each of these documents provides a different perspective on what is meant by the term. Finally, all of these definitions are further confounded by debates about subject matter and confusion about whether these terms refer to only what is governed by regulations, what is governed by any written guidelines and standards, the actual practice of researchers, normative or ideal practices, or the content of an appropriate curriculum. These challenges are only amplified when transferred to the international setting.

While there has been much less focus on regulation and education in this area internationally than in the USA, the relevant issues are widely accepted as important (e.g. Ham, 2010; ESF, 2010). Unfortunately, all of the challenges described above within the USA are further confounded by the context of different cultures, different research environments, different languages, and different understandings of what is meant in other languages. Clearly it cannot be assumed that any particular choice of words will be understood in the same way by different audiences, so it is best to define how the words will be used here.

For the purpose of this discussion, the term "research ethics" will be used synonymously with "responsible conduct of research", "RCR", and "research integrity". Research, ethics, and research ethics will be defined here as follows:

- *Research*: All scholarly activities, including but not limited to observation, discovery, experimental study, development, and use of new knowledge.

- *Ethics*: Reflection about how best to act, and how to make choices in the face of competing interests, principles or outcomes.
- *Research ethics*: Examination and understanding of the practices and responsibilities of all of those who have a role in research, the corresponding written regulations, guidelines, and policies, the unwritten standards that govern the practice of research, and the challenges that arise when standards may not be clear or known.

Because discussions about research ethics overwhelmingly begin with a focus on "research misconduct", it is important to consider the definition of this frequently used term. Research misconduct is defined under US regulations as "... fabrication, falsification, or plagiarism in proposing, performing, or reviewing research, or in reporting research results" (OSTP, 2000); however, it is fair to note that the domain of concern can be more generally defined to include intentional, serious misrepresentations that corrupt the research process or record. By either definition, everyone can agree that such actions are unacceptable and have no place in the research environment. On the other hand, it is worth asking whether "research misconduct" is really the real problem to be addressed or whether making it the primary focus is not in fact the best way to proceed.

2 THE PROBLEM OF RESEARCH MISCONDUCT

One of the problems with a focus on research misconduct is what it says to the people who are required to participate in research ethics education programs. The message in such cases risks suggesting that the audience either (a) does not yet know that it is wrong to lie, cheat, or steal, or worse; (b) does know that it is wrong to lie, cheat, or steal, but chooses to do so anyway. Neither of these stances is a good place to start for the education of adults.

Another possibility is that research misconduct is in fact frequent, perhaps even characteristic of the research enterprise. Certainly this seems to be the implication of several recent reports (Martinson *et al.*, 2005; Titus *et al.*, 2008). If this is in fact the case, then it brings to mind the possibility that we are faced with more fundamental problems than research ethics education can resolve. If research misconduct is in fact so pervasive, then this surely is a sign that the problem extends well beyond the boundaries of the research environment. If so, it is probably not reasonable to expect that a course in research ethics will somehow supplant what is widely accepted in society; instead, the risk is high that emerging researchers would cynically view a focus on ethics as misguided and out of touch. However, before concluding that research misconduct occurs so widely, it is worth considering the possibility that research misconduct is neither as pervasive nor as damaging as might at first be feared.

If research misconduct were so pervasive, then it seems implausible that we could have seen the rapid advances and successes that have been the hallmarks of science and technology in just the most recent 50 years. If anything, some have argued that the pace of scientific progress is too fast, running too far ahead of our ability to provide the necessary social and legal framework to accommodate marked changes in our world. While the frequency and the damage caused by research misconduct may be less severe than implied by some recent research studies, this is not to say that we should condone research misconduct. Whether research misconduct is committed by 1% or 30% of scientists, the percentage is clearly non-zero and it would be better if it could be reduced.

If the above observations suggest that education about research misconduct may be less than productive (and possibly counterproductive), and if the only goal of research ethics education is to decrease research misconduct, then it might appear that there is no point to such education. However, this assumes that the fundamental problem is that research misconduct occurs only because flawed individuals make bad decisions. Reframing the question highlights another — perhaps more useful — way to think about this issue. Rather than viewing research misconduct in isolation, it is worth considering the more systemic approach now used successfully, for example, when errors or failures occur in the transportation and medical industries (e.g. Dekker, 1997; Reason, 1990; IOM, 2002). Rather than viewing research misconduct as a disease, it might be useful to think of it as a symptom instead.

3 WHY FOCUS ON RESPONSIBLE CONDUCT OF RESEARCH?

If research misconduct is viewed as a symptom, then the focus shifts from solving the problem of research misconduct to instead identifying the factors that allow research misconduct to occur. From this perspective, it is interesting to note the extent to which nearly every known case of research misconduct is characterised by failures not so much of individual moral fiber but of the responsible conduct of research. Table 1 lists examples of such failures that are typically characteristic of, and *necessary* for, most cases of research misconduct.

Two characteristics of the items listed in Table 1 should be noted. First, none of these items is rooted in deliberate, serious misrepresentations by scientists; all simply reflect the ways in which scientists conduct research and interact with one another. Second, while these are demonstrably all important in ensuring the integrity of research, many of these are dimensions of the practice of research that very likely will not have been learned in kindergarten, and probably not in college. In short, a focus on these areas is likely to be important for not only promoting *responsible* conduct of research, but also creating an environment in which it much more difficult and much less likely that research misconduct will occur. This argument is consistent with the finding, for example, that "perceptions of fair treatment in the work environment appear to play important roles in fostering — or undermining — research integrity" (Martinson *et al.*, 2010).

4 WHY TEACH ABOUT RESEARCH INTEGRITY?

The first question to be asked about the teaching of research integrity should certainly be: Why teach about research integrity? Without clarity about the goals of teaching, it makes little sense to begin designing a program or approach. Based on the above discussion of research misconduct, it seems reasonable to argue that the goal of such teaching should not be limited to decreasing research misconduct. And although an increasing focus on requirements for such education are found not only in the USA, but other countries as well (e.g. DFG, 1998; SCJ, 2006; NHMRC/AVCC, 2007; Zeng and Resnik, 2009), it would be cynical and probably counterproductive to teach research ethics only because we are required to do so. Unfortunately, there is no simple answer to the question of what goals should be pursued.

Kalichman and Plemmons (2007) addressed the question of goals for teaching research ethics by reporting on interviews of 50 teachers of RCR courses in the USA. The findings were remarkable for the diversity of responses and not for any sense of common, shared views of what we are or should be trying to accomplish. Table 2 briefly summarises examples of the wide range of goals reported. It is noteworthy that these goals have many

Table 1. "Necessary Failures" for research misconduct to occur (some of this text is adapted from Kalichman and Plemmons, 2009).

Necessary Failure	Description of Responsibility	Consequences of Failure
Recordkeeping	Researchers should keep records sufficient to reconstruct the work that has been done.	Easier for both intentional and unintentional misrepresentations to occur.
Managing Bias	Researchers should remove, manage, or mitigate the risks of bias in their research.	Increased risk of both intentional and unintentional bias in the conduct or reporting of research.
Collaboration	Researchers should encourage openness in inquiry, questioning, discussing, and sharing among collaborators.	Increased risk of misunderstandings and decreased sense of personal accountability to others.
Authorship	Researchers should give credit where credit is due and accept credit only when warranted.	Increased risk that co-authors will be insufficiently aware of the integrity of the work being published.
Peer Review	Peer reviewers should provide a critical and constructive review of work submitted for their consideration.	Increased risk that fundamentally flawed, plagiarised, or misrepresented work will be accepted for funding or publication.
Asking Questions	If in doubt, researchers should ask for clarification.	Increased risk that unintentional, or even intentional, misrepresentations or failings will occur in the conduct or reporting of research.
Whistleblowing	Researchers who believe that they have witnessed misconduct should act.	Increased risk that the research record will be corrupted, that the public's perception of science will be tarnished, and that the witness will later need to answer allegations of complicity if misconduct is reported by someone else.

(*Continued*)

Table 1. (*Continued*)

Necessary Failure	Description of Responsibility	Consequences of Failure
Mentoring	Researchers should serve as mentors to those who are following in their footsteps, and should seek the mentoring they need for their own path.	Increased risk that researchers will lack the necessary understanding of standards of responsible conduct and of how to deal with the inevitable ethical challenges that are part of the conduct of research.

Table 2. Goals for teaching research ethics (adapted from Kalichman and Plemmons, 2007).

Goal	Description	Examples
Knowledge	What would it be helpful for learners to know?	Ethical decision-making frameworks and theories, rules and regulations, methods of science, resources for further information
Skills	What would it be helpful for learners to be able to do?	Ethical decision-making, problem solving, communication
Attitudes	What perspectives and beliefs would it be helpful for learners to hold?	Importance of research ethics, personal responsibility for promoting integrity in the research environment
Behaviour	What actions is it hoped that learners will carry out?	Adherence to high standards in the conduct of research, effective communication with others about research ethics, not committing research misconduct

different qualities ranging from simple knowledge (e.g. where would you find information about guidelines for authorship?) to behaviours (e.g. honestly report research findings), that nearly all of these goals are desirable outcomes, and that it is unrealistic to expect that any one course would be able to successfully address everything envisioned by this list.

Since not all of these goals are achievable in the context of a single educational intervention, it would be valuable for any instructor to begin by distilling her or his goals down to what they see as the core purpose of such education (e.g. Kalichman, 2007).

5 WHO SHOULD BE TAUGHT?

Part of the responsibility for an educator is to determine not only why teach, but who should be taught. Many in academia see research ethics education as most appropriate for graduate students, those who are starting to learn about how to do research. This makes at least some sense, but there are several reasons to consider the audience to be much wider (Bulger and Heitman, 2007). First, there are many different dimensions to the topic of research ethics and some are better covered at earlier stages (e.g. for undergraduate students), while others may be more appropriate for later in a research career (e.g. for postdoctorates or faculty). And even within these groups, there are reasons to expect different needs at different stages (e.g. new versus senior graduate students). Second, general education varies in different disciplines, institutions, and even countries. As a result it cannot be assumed, for example, that all postdoctorates have had sufficient or appropriate education in research ethics. In short, given that the scope of research ethics includes roles and responsibilities of all of those engaged, the potential audiences for instruction are wide ranging (Table 3).

Table 3. Potential audiences for research ethics education.

Graduate Students
Postdoctorates
Undergraduate Students
Staff, Administrators
Ethics Committee Members
Faculty (New, Senior)

6 WHAT SHOULD BE TAUGHT?

Once the question of who should be taught has been answered, the next question is *what* should be taught. Just as there are many different possible goals for research ethics education, there is a limitless range of possible

Table 4. Topics for research ethics courses.

General Topics	Specific Topics and Disciplines
Data Management	Human Subjects
Conflict of Interest and Commitment	Animal Subjects
Collaboration	Stem Cells
Authorship	Dual Use Technology
Publication	Environmental Protection
Peer Review	Computational Biology
Mentoring	Computers and Information Technology
Social Responsibility	High Energy Physics
Research Misconduct	Cultural Anthropology
Questionable Research Practices	
Asking Questions	
Dispute Resolution	
Whistleblowing	

topics. Different lists of topics have been proposed even within a single US agency, the Department of Health and Human Services (NIH, 1989; PHS, 2000; NIH, 2009), and the coverage of topics varies widely in some of the most widely used US texts for teaching research ethics (Bulger *et al.*, 2002; Macrina, 2005; National Academies, 2009). Table 4 provides two sample lists of such topics. The first list covers items that are likely to be relevant to almost anyone involved in research, while the second covers examples of areas which are likely to have very specific issues that are relevant to only some research disciplines.

7 HOW SHOULD RESEARCH INTEGRITY BE TAUGHT?

In the 1960s, Marshall McLuhan coined the phrase "the medium is the message" (McLuhan and Lapham, 1964). While the meaning and applicability of this phrase has been widely discussed, it seems particularly apt for the topic of research integrity. Even more than questions of content, the question that lingers is whether "ethics" or "integrity" is something that can or should be taught. There is a high risk that instead of conveying the message that this subject is, in fact, important, a poorly taught or inadequate approach to research ethics can be worse than having done nothing at all. Clearly, the message conveyed by an instructor who is not knowledgeable and interested in his or her topic, or by simply requiring a pre-packaged

Web-based tutorial, is that the issues raised are neither important nor substantive. For these reasons, it is essential to consider where such material should be taught and what methods of teaching are likely to best meet the intended goals of research integrity education.

The teaching of research ethics might occur in many different "places." The principal options include the classroom, on the Internet, and in the research environment itself. For the sake of this overview, brief descriptions are provided for each of these settings in Table 5. Although there are multiple options for settings, these options should not be seen as mutually exclusive. It is likely that the best choice is some combination of these possibilities. For example, web-based instruction can be an excellent adjunct

Table 5. Where should research ethics be taught?

Setting	Description
Classroom: Course	Full course, taught in classroom, dedicated to topic of research ethics; most appropriate for undergraduate or graduate students, but could also be created for other groups of researchers; typically includes both lectures and discussions to engage students in addressing the challenges of research ethics. Example: *Scientific Ethics*, UC San Diego http://ethics.ucsd.edu/courses/ethics
Classroom: Across the Curriculum	Insertion of ethics curriculum into curriculum of discipline-specific courses for undergraduate or graduate students; cumulative focus of a single research ethics course is lost with this format, but it provides a demonstration that research ethics is part of the curriculum rather than something to be addressed separately. Resource: *Society for Ethics Across the Curriculum* http://www.rit.edu/cla/ethics/seac
Internet: Course	Full research ethics course taught on the Internet distinguished in two ways from a classroom course: (1) online readings (or video lectures) instead of in person classroom lectures and (2) discussion by electronic messaging (e.g. email, discussion forums, chat rooms) rather than in person. Example: *Scientific Integrity*, UC San Diego http://ethics.ucsd.edu/courses/integrity

(*Continued*)

Table 5. (*Continued*)

Setting	Description
Internet: Tutorial	Content of a research ethics course can be surveyed in an online tutorial, which provides information to the learner, and may include questions to assess understanding of what has been learned, but does not require active engagement through discussion with other learners or an instructor. Example: *Responsible Conduct of Research*, CITI https://www.citiprogram.org/rcrpage.asp
Research: In context	Introduction of research ethics into many different possible forums, including group meetings, journal clubs, research lecture series, or even one-on-one mentoring; the intended message is that addressing ethics is clearly a part of research. Resource: *Research Context*, Resources for Research Ethics Education http://research-ethics.net/educational-settings/research-context

and starting point for subsequent instruction in the classroom or in the research context.

The question of methods for teaching in this area is one that has at least a partial answer. Nearly everyone who has spent time teaching research ethics or written about the topic has recognised that the enjoyment, and presumably the impact, of such teaching is greatly enhanced by active learning approaches rather than by lectures alone (e.g. Jones *et al.*, 2010). This is not surprising. The evidence is compelling that engaging learners in "creating" their own learning experience, i.e. "active learning", has numerous advantages over the passive experience of simply receiving information through a lecture or reading material (Lambert and McCombs, 1998; Bransford *et al.*, 1999). One of the most frequently cited examples of how best to teach research ethics is through the use of case studies (e.g. Reiser and Heitman, 1993; Macrina and Munro, 1995; Bebeau *et al.*, 1995). Learners are expected to be actively engaged by struggling with questions about real or contrived scenarios that do not have easy answers.

The use of case studies for teaching research ethics is certainly valuable, but it is by no means the only option for producing a more engaged, active learning environment (Table 6). As for choosing instruction settings, it is

Table 6. Examples of tools for encouraging research ethics discussion. Detail and examples for the following can be found on the Resources for Research Ethics Education website at: http://research-ethics.net.

Approach	Description
Case Studies	Learners are given a short story or scenario describing a challenge in research ethics; typically the task is for a small group of participants to identify the challenges, propose solutions, and seek agreement or at least understanding of different positions.
Role Playing	Learners are given a description of a challenging scenario involving two or more participants and assigned to play the roles of those involved; their task is to attempt resolution of the challenges from the perspective of their assigned roles.
Debates	Individuals or small groups are assigned to take opposing viewpoints on a specific question or situation, to make their respective cases for (or against), and address the arguments made from opposing perspectives.
Email Discussion	Similar to case studies in a classroom setting, small groups are assigned to address cases or questions through electronic discussion (e.g. by email, discussion forums, or chat rooms).
Video	Instead of simply reading a case study, learners watch either a short video or even a full length movie in which actors struggle with the ethical challenges of conduct of research, and then discuss or react to what they have seen.
Question-based lectures	Rather than simply giving a lecture, the instructor asks a series of questions that challenge learners to develop problem-solving strategies and to ask for information that they perceive is needed; this is supplemented by additional information from the instructor to fill in gaps or correct misunderstandings.
Surveys	In advance of meeting or discussion, participants are asked to complete a brief questionnaire highlighting different perceptions on selected research ethics issues; survey results are then used as a prompt for class discussion of the meaning of diverse responses.

(*Continued*)

Table 6. (*Continued*)

Approach	Description
Journal Publications	Most research disciplines will have current journal articles that can be used to promote ethics discussion either directly (e.g. a published report about the handling of conflicts of interest by researchers in the particular discipline) or indirectly (e.g. questions could be asked about the meaning of authorship in the case of a paper with over 50 authors).
Literature	Well-written biographies, histories, novels, short stories, and even poetry provide learners with an insightful look into a richer view of the challenges of research than might be conveyed by a brief case study.
Current Events	Students are assigned to find and briefly report on recent stories reflecting on the ethical dimensions of the conduct of science as reported in either the popular or scientific press; these stories can be used as a point of departure to discuss any of the topics typically covered in research ethics courses.
Student Teaching	With guidance from the course instructor on content and methods, individual students or small groups are each given the responsibility for teaching a selected research ethics topic to other course participants.
Guest Faculty	Guest faculty can participate individually or as part of a panel to provide perspectives from differing disciplines.
Textbooks	An entire course can be structured around any one of numerous textbooks written on the topic of research ethics.

plausible that the best answer will typically be to use more than one of these options.

8 NEXT STEPS?

While the importance of a focus on evidence-based research ethics (Steneck and Bulger, 2007; Kalichman, 2009) is clear, it is not as clear how the challenge of education should be best met. It is certainly possible that many possible approaches are valid and even exemplary, and it is also possible that what is best in one place or circumstance is less than ideal for a different place or circumstance. Finding common ground for diverse individuals, research groups, and disciplines across different cultures and nations is a formidable challenge. Such a comprehensive agreement may in fact be unrealistic, but it would certainly be worth exploring the following four questions about goals, audiences, settings, tools, and topics for research ethics education:

- On what points can we find multi-national agreement?
- On what points do we find significant agreement?
- What are the evidence-based, best available, or at least most promising practices?
- What mechanisms would best serve for creating or facilitating multi-national collaborations to address areas of common interest?

References

1. Bebeau, M. J., Pimple K. D., Muskavitch, K. M. T., Borden, S. L. and Smith, D. H. (1995). *Moral Reasoning in Scientific Research: Cases for Teaching and Assessment.* Bloomington, IN: Poynter Center for the Study of Ethics and American Institutions. Accessed from http://poynter.indiana.edu/mr/mr.pdf.
2. Bransford, J. D., Brown, A. L., Cocking, R. R. (eds.) (1999). *How People Learn: Brain, Mind, Experience, and School.* Washington, DC: National Academy Press.
3. Bulger, R. E. and Heitman, E. (2007). Expanding responsible conduct of research instruction across the university. *Academic Medicine*, 82: 876–878.
4. Bulger, R. E., Heitman, E. and Reiser, S. J. (2002). *The Ethical Dimensions of the Biological Sciences*, 2nd edition. New York, NY: Cambridge University Press.
5. Dekker, S. (1997). *The Field Guide to Understanding Human Error.* Burlington, VT: Ashgate Publishing.
6. DFG (1998). Proposals for safeguarding good scientific practice. *Recommendations of the Commission on Professional Self Regulation in Science, Deutsche Forschungsgemeinschaft.* Accessed from http://www.dfg.de/download/pdf/foerderung/rechtliche_rahmenbedingungen/gute_wissenschaftliche_praxis/self_regulation_98.pdf.
7. DHHS (2001). Notice of suspension of "PHS Policy on Instruction in the Responsible Conduct of Research." Federal Register, 66(35):

11032–11033. Accessed from http://ori.dhhs.gov/education/ congressional_concerns_ published.shtml.
8. Dresser, R. (2001). *When Science Offers Salvation: Patient Advocacy and Research Ethics.* New York, NY: Oxford University Press.
9. ESF (2010). *Fostering Research Integrity in Europe. Executive Report.* A report by the ESF Member Organisation Forum on Research Integrity, June 2010. Accessed from http://ori.hhs.gov/blog/wp-content/uploads/ 2010/07/research_ integrity_exreport.pdf.
10. Ham, B. (2010). CAST and AAAS Announce Joint Steering Committee on Ethics in Science. *News Archives, Science.* Accessed from http://www.aaas.org/news/ releases/2010/0630cast_statement.shtml.
11. Iltis, A. S. (ed.) (2006). *Research Ethics.* New York, NY: Routledge.
12. IOM (2002). *Integrity in Scientific Research: Creating an Environment That Promotes Responsible Conduct.* Washington, DC: National Academies Press. Accessed from http://www.nap.edu/catalog.php?record_id=10430.
13. Israel, M. and Hay, I. (2006). *Research Ethics for Social Scientists.* London: Sage Publications.
14. Jones, N. L., Peiffer, A. M., Lambros, A., Guthold, M., Johnson, A. D.,Tytell, M., Ronca, A. E., and Eldridge, J. C. (2010). Developing a problem-based learning (PBL) curriculum for professionalism and scientific integrity training for biomedical graduate students. *Journal of Medical Ethics* [Epub ahead of print]. Accessed from http://www.ncbi.nlm.nih.gov/pubmed/20797979.
15. Kalichman, M. W. (2007). Responding to challenges in educating for the responsible conduct of research. *Academic Medicine*, 82(9): 870–875.
16. Kalichman, M. W. (2009). Evidence-based research ethics. *American Journal of Bioethics*, 9: 85–87.
17. Kalichman, M. W. and Plemmons, D. K. (2007). Reported goals for responsible conduct of research courses. *Academic Medicine*, 82(9): 846–852.
18. Kalichman, M. S. and Plemmons, D. K. (2009). Overview. In *Topics, Resources for Research Ethics Education.* Accessed from http://research-ethics.net/topics/ overview.
19. Lambert, B. L. and McCombs, N. M. (eds.) (1998). *How Students Learn: Reforming Schools through Learner-Centered Education.* Washington, DC: American Psychological Association.
20. Macrina, F. L. (2005). *Scientific Integrity: An Introductory Text with Cases.* 3rd edition. Washington, DC: American Society for Microbiology Press.

21. Macrina, F. L. and Munro, C. L. (1995). The case-study approach to teaching scientific integrity in nursing and the biomedical sciences. *Journal of Professional Nursing*, 11(1): 40–44.
22. Martinson, B. C., Anderson, M. S. and De Vries, R. (2005). Scientists behaving badly. *Nature*, 435(7043): 737–738.
23. Martinson, B. C., Crain, A. L., De Vries, R. and Anderson, M. S. (2010). The importance of organizational justice in ensuring research integrity. *Journal of Empirical Research on Human Research Ethics*, 5(3): 67–83.
24. McLuhan, M. and Lapham, L. H. (1964). *Understanding Media: The Extensions of Man.* Cambridge, MA: MIT Press.
25. National Academies (2009). *On Being a Scientist: A Guide to Responsible Conduct in Research.* National Academy of Sciences, National Academy of Engineering, and Institute of Medicine. Washington, DC: National Academies Press. Accessed from http://www.nap.edu/catalog.php?record_id=12192.
26. NHMRC/AVCC (2007). *Australian Code for the Responsible Conduct of Research.* Accessed from http://www.nhmrc.gov.au/_files_nhmrc/file/publications/ synopses/r39.pdf.
27. NIH (1989). *Requirements for Programs on the Responsible Conduct of Research in National Research Service Award Institutional Training Programs.* Guide for Grants and Contracts on December 22, 1989, 18(45): 1. Accessed from http://grants.nih.gov/grants/guide/historical/1989_12_22_Vol_18_ No_45.pdf.
28. NIH (2009). Update on the Requirement for Instruction in the *Responsible Conduct of Research.* Notice Number NOT-OD-10-019. Accessed from http://grants.nih.gov/grants/guide/notice-files/NOT-OD-10-019.html.
29. NSF (2010). B. *Responsible Conduct of Research in Grantee Standards.* Accessed from http://www.nsf.gov/pubs/policydocs/pappguide/nsf10_1/aag_4.jsp#IVB.
30. OSTP (2000). *Federal Research Misconduct Policy.* Office of Science and Technology Policy in the Federal Register, 65(235): 76260–76264. Accessed from http://ori.hhs.gov/policies/fed_research_misconduct.shtml.
31. PHS (2000). *PHS Policy on Instruction in the Responsible Conduct of Research (RCR).* Accessed from http://ori.dhhs.gov/policies/RCR_Policy.shtml.
32. Reason, J. (1990). *Human Error.* Cambridge, UK: Cambridge University Press.
33. Reiser, S. J. and Heitman, E. (1993). Creating a course on ethics in the biological sciences. *Academic Medicine*, 68(12): 876–879.

34. SCJ (2006). *Statement: Code of Conduct for Scientists. Science Council of Japan.* Accessed from (English version) http://www.scj.go.jp/ja/info/kohyo/pdf/kohyo-20-s3e-1.pdf (Japanese version). http://www.scj.go.jp/ja/info/kohyo/pdf/kohyo-20-s3.pdf.
35. Steneck, N. H. and Bulger, R. E. (2007). The history, purpose, and future of instruction in the responsible conduct of research. *Academic Medicine*, 82(9): 829–834.
36. Titus, S. L., Wells, J. A. and Rhoades, L. J. (2008). Repairing research integrity. *Nature*, 453(7198): 980–982.
37. Zeng, W. and Resnik, D. (2009). Research integrity in China: Problems and Prospects. *Developing World Bioethics.* DOI 10.1111/j.1471-8847.2009.00263.x. Accessed from http://onlinelibrary.wiley.com/doi/10.1111/j.1471-8847. 2009.00263.x/pdf.

CHAPTER 30

ESTABLISHING AN INSTITUTIONAL CULTURE OF RESEARCH INTEGRITY: KEY CHALLENGES & SUCCESSFUL SOLUTIONS

Philip J Langlais

Research integrity and adherence to professional standards of conduct are critical to the goals of our educational systems and to the well-being of our national and global societies. A growing body of literature and research suggest that attaining these goals will require not just an educational program but the establishment of a culture and environment that promotes and rewards research integrity and a sense of social responsibility.

Establishing a stronger institutional culture of research integrity has its challenges. Fortunately, there are strategies for addressing those challenges. The following is a brief description of some of these challenges and strategies that are based on my experiences as a Vice Provost, member of the Council of Graduate Schools Projects on Responsible Conduct of Research, and interactions with fellow graduate deans and vice-presidents for research at several U.S. universities. It is important to realise that each institution, administration, and discipline has its own unique culture and set of norms to which these challenges and solutions may not apply equally.

"Buy-in" from faculty and upper leadership is essential to a strong and enduring culture of research integrity. Obtaining buy-in, however, is difficult when faculty and administrators have the following attitudes and perceptions:

- Misconduct and misbehaviors "rarely", if ever, occur on this campus
- Faculty members are providing more than adequate RCR training to students
- RCR training applies only to biological and medical research

- RCR training is required only to meeting federal human and animal subjects regulations
- Faculty workloads are already excessive, leaving no time to put more effort into RCR training or mentoring
- RCR training requires major changes to curriculum and significant new resources; there is no room in current curriculum
- Faculty training and mentoring of students in research ethics and practices are personal and private, and "outside interference" violates academic freedom

ODU and other US universities have to varying degrees successfully addressed these obstacles by implementing an assessment and educational campaign led by top administrators and faculty champions. These coordinated "top down" and "bottom up" approaches emphasise honesty, integrity, respect and social responsibility as core institutional values and promote a culture of research and scholarly integrity "as the right thing to do".

Successful programs have been directed by a university committee comprised of top university officials, e.g. Vice Provost, VP of Research, Dean of Graduate School, and a number of highly respected and published faculty who are strongly committed to enhancing RCR training and research and scholarly integrity. Other committee members have included the institutional compliance officer, director of student judicial affairs, and students.

Assessing the attitudes and perceived effectiveness of current RCR training is essential to a successful campaign. In the case of ODU, the university RCR committee developed an initial survey to assess perceptions of value and the importance of RCR training among graduate students, faculty and administrators. A second survey administered two years later assessed the effects of initial training programs on attitudes and perceptions, and asked additional questions about the nature and frequency of training in several areas of research conduct and integrity. The results demonstrated that while most of the respondents considered RCR training and research ethics as very important, far more faculty than students judged RCR training as adequate. Depending on the RCR topic, e.g. conflict of interest, data management, between thirty to fifty percent of the students indicated that they had received no training. These findings were presented to the university community together with research findings on the incidence and impact of misconduct and serious misbehaviours. As a result, faculty awareness and support for RCR education has increased and students have exerted pressure on faculty, department chairs, and university administrators to provide better training.

Based on the assessments, the RCR committee constructed a university plan outlining specific areas and methods for incorporating RCR training and enhancing the culture of research integrity. In response to the committee's plan and recommendations, several departments and graduate programs have voluntarily incorporated RCR topics into existing required courses such as research design and methods, and an introduction to scientific inquiry. Certain departments/schools have encouraged faculty members to voluntarily adopt thesis and dissertation agreements on authorship and use of data. Others have elected to include questions on research integrity and social responsibility/impact on comprehensive exams and advancement to candidacy exams. Some department chairs and associate deans have added an entire lecture or a brief presentation on research integrity and professional standards in their orientation programs for new undergraduate and graduate students.

ODU and other universities have allocated resources to provide better training and preparation of junior faculty and those graduate students who will soon enter universities and research centers. By incorporating RCR training and topics into new faculty professional development and preparing future faculty (PFF) programs, institutions demonstrate a long-term commitment to RCR and scholarly integrity. Recognition by the President, Provost or Dean of Faculty, of postdoctorates and students for their outstanding contributions to research and scholarly integrity sends a strong message to the community that these behaviours are valued and rewarded by the upper administration. Several institutions have also established Outstanding Mentor awards to promote and recognise faculty who exemplify the highest professional standards, teaching of ethics and mentoring skills. A major review of university policies by upper administration, faculty, university counsel, university auditors, and students can produce significant and positive changes in the institutional culture of research integrity. Among the policies and processes with measureable effects on research integrity are conflicts of interest, investigation of and response to allegations of infractions of RCR standards and ethical behaviour, whistleblowing and protection from retaliation, criteria for awarding faculty tenure and promotion, and assignment of faculty workloads.

In conclusion, establishing a stronger institutional culture of research integrity has its challenges. Fortunately, there are effective strategies that can be used to address those challenges. Strong leadership from faculty and administration along with persistence and patience are essential to achieving measureable and enduring changes in institutional culture.

CHAPTER 31

A CURRICULUM FOR RCR TRAINING IN GERMANY[1]

Gerlinde Sponholz

The "Curriculum 'Good Scientific Practice' for Courses in Science and Medicine" was developed on behalf and in cooperation with the ombudsman of the Deutsche Forschungsgemeinschaft DFG (German Research Foundation).[2]

1 BACKGROUND

In 1997, after a severe case of misconduct in Germany, the executive board of the DFG[3] had appointed an international commission to develop recommendations to foster and safeguard good scientific practices (GSP). Since 1998, all universities or other research institutions in Germany have had to set up guidelines for safeguarding GSP and rules for handling allegations of suspicious cases of misconduct, as well as to establish ombudspeople and commissions for investigations. These are obligations for all institutions to get financial support from DFG. But training in GSP is not required.

[1]Office of the Ombudsman for Science (Ombudsman für die Wissenschaft): Prof. Ulrike Beisiegel, Prof. Dr. Siegfried Hunklinger, Prof. Dr. Wolfgang Löwer, Helga Nolte, Hendrik Plagmann. Didactics, ideas, discussion: Jutta Baitsch, Gerhard Fuchs, Martina Geiselhart, Sabine Just, Prof. Dr. Frieder Keller, Prof. Dr. Gerd Richter. English translation: Prof. Dr. Josef Leidenfrost.

[2]The name was changed in 2010 into Ombudsman für die Wissenschaft. Accessed from http:// www.ombuds-wissenschaft.de.

[3]The second part of the document is written in English: Deutsche Forschungsgemeinschaft (1998) Sicherung guter wissenschaftlicher Praxis. Safeguarding Good Scientific Practice. Denkschrift. Weinheim: Wiley-VCH. Accessed from http://www.dfg.de/ antragstellung/gwp/index.html.

In Germany, an explicit education in GSP is still voluntary and only a few scientists have developed and offered teaching programs in GSP.

Inside the regular meetings of ombudspeople (they are organised by the Ombudsman für die Wissenschaft), there were intensive discussions about education in GSP; special workshops about this topic had been organised. There was a common experience: only few students/ scientists knew about the DFG recommendations or the guidelines of their universities and the existing ombudsman system in Germany. Scientific integrity is not a regular matter for studies in German universities even for postgraduates. Many ombudsmen have asked for a curriculum for GSP training and wanted support for teaching materials and for ideas to develop their own education programs.

In 2009, the curriculum was finished and is now available in German and soon in English.[4] The intention to develop this curriculum was to give support to all scientists/teachers in universities or research institutions to establish their own teaching program for GSP, and to foster an ongoing discussion about the complex field of scientific integrity.

2 STRUCTURE OF THE CURRICULUM

The history and background for GSP education in Germany and the USA is described in the first part. The second part is about a theoretical framework for didactics, theory of development and training of competences (as important elements of the responsible professional conduct for scientists) and about teachers' qualifications.

Suggestions for concrete realisation are presented in the third part. We recommend at least two opportunities to learn about and discuss the topics of research integrity for undergraduate and postgraduate students in science and medicine.

The first step should be an introduction into GSP, some important cases of scientific misconduct, and how they were manged. This could be organised in a two-hour lecture with discussions. It should be available for all students at the beginning (science) or middle (medicine) of their studies. The teachers should have a honest interest and, if possible, a training in GSP. The ombudspeople of the organisations should be involved.

[4]Download: http://www.ombuds-wissenschaft.de.

The second opportunity should be a course or workshop with a minimum of 14 hours; students (in medicine) or postgraduate students should attend the course at the beginning of their doctorate. The group size should be between 12 to 20 participants. The teaching style should be interactive with case discussions, role playing, discussions of one's own experience and problems, information about and discussion of rules, regulations and guidelines. We developed modules that could be shaped depending on the needs and the participants. The modules we suggest are:

- Introduction to Scientific Integrity
- Scientific Misconduct
- Data Management
- Publication Process and Authorship
- Mentor and Trainee Responsibilities
- Research with Human Subjects
- Research with Animals
- Conflicts of Interest and Scientific Cooperation
- Managing Conflicts

For each module, learning objectives, contents and didactic hints are formulated. Teaching materials are available as downloads on the website from the Ombudsman für die Wissenschaft.

Part four of the curriculum is about the unsolved problem of evaluation and the further development of the curriculum. There is a need to adapt the materials and perhaps some contents of the curriculum to other fields of research, especially to humanities. New international standards of education in GSP should be included.

3 SOME PROBLEMS, NEEDS AND INITIAL EXPERIENCES

In Germany, we have no requirement for developing and attending courses in GSP. Because of that, it is often difficult to implement activities in GSP into teaching programs of universities. Since fall 2009, an increasing number of scientist and especially managers of graduate schools have asked for lectures and courses, but there are only few teachers/scientists with a training in GSP in Germany. We have not yet developed a German or European network for education in GSP, for teachers' training courses or for the development of appropriate evaluation materials.

Initial feedback from more than 10 lectures and courses (since fall 2009) based on the curriculum suggest that the main problems with this type of training are mostly technical ones (information about the lectures/courses, time problems especially with weekend courses). Different expectations on the part of participants can raise additional difficulties. Other problems include the gap between the norms, rules and the regulations which underlay GSP and the reality of scientific world with the real conflicts that arise in situations, low levels of commitment from faculty members and advisors, and problems of competition and pressure.

If there is adequate support from faculty members, all of the recommended lectures and courses could be easily implemented. The participants enjoyed working with concrete cases and liked comparing them with the regulations of their universities. Most of the participants proposed that lectures and courses should be obligatory for all undergraduate, postgraduate students, and advisors.

CHAPTER 32

TEACHING AND TRAINING RESEARCH ETHICS

Ragnvald Kalleberg

I shall reflect on two general challenges in teaching and training, on how to foster (1) an understanding in scientific communities of the internal connection between epistemology and ethics; and (2) an understanding that research ethics is primarily an institutional responsibility.

1 RESEARCH ETHICS: EXTERNAL NORMS OR INTEGRAL PART OF RESEARCH?

There is a tendency in some scientific communities to understand research ethics as external to the scientific enterprise, something imposed by external bodies. This view is often combined with a scientistic theory of science, according to which only documentation and empirical analysis can be rational. Normative and ethical disagreements cannot be settled with the force of better arguments.

Such a view is not tenable. Epistemological and ethical components are interwoven in research practice. Unavoidable values in well-functioning research fields are, for instance, truth, consistency and impartiality. Norms are internal to science. Norms are not irrational conventions based on emotions and conventions. When we claim that it is wrong to fabricate data or plagiarise, we do not just express feelings or adhere to contingent conventions. We presuppose that the normative arguments are valid, and have a basis in the research system itself.

According to my experience from PhD courses, the most efficient way to convey an understanding of the internal connection between epistemology and ethics is by focusing on what doctoral students are doing in their own research. Then they may learn to see how such norms and values are relevant to their own research practice. They may also experience that ethical criticism, for instance about irrelevance of topics or improper citations, regularly contribute to more interesting research questions and precision. Moral reflections can improve the research process and the resulting output of knowledge.

The persons primarily responsible for teaching and training science ethics should be located in the discipline or field in question. In some fields of natural science, there has been a tendency only to rely on ethicists from the humanities. That does not promote a realistic understanding of the importance of science ethics in all phases of a project. It may instead stimulate a conception of ethics as something external to the field.

Obviously, there are other, more indirect ways of conveying an understanding of the internal relationships between epistemology and ethics, such as focusing on social theory and theory of science (Boudon, 2001; Habermas, 2003; Kalleberg, 2007) and/or using historical and sociological studies about the emergence and development of science. The American sociologist Robert Merton documented, for instance, how modern science from the scientific revolution was also an ethical project. A catchword is the conception of an "ethos of science", referring to ethical norms prescribing, proscribing, permitting and reccomending scientific conduct (Merton, 1973, Part 3). Naturally, when Merton argued this he did not invent research ethics, he explicated informal norms that had been at work for three centuries.

2 TEACHING RESEARCH ETHICS: ONLY FOR INDIVIDUALS, NOT COLLECTIVES?

There is a tendency to focus only on individuals when discussing research ethics. This is inadequate, both as a theoretical perspective and as a preparation for action. Research ethics is an institutional phenomenon, referring to a complex set of values and norms, embedded in practices and attitudes on different levels, such as individuals, research groups, disciplines, institutions, national and international organisations. Individual scientific conscience is real, but only one element in a larger picture.

One way of reminding ourselves about this basic institutional fact is to analyse situations where essential norms and values have been violated. There are many to choose from, such as medical research in Germany or Japan during the 1930s and 1940s, the infamous Tuskegee project in the

USA from the 1930s to 1970s, or the recent reports about problems in Chinese universities (see Walker, 1995; Harris, 1992; Jones, 1993; Wang, 2006; the Lancet 2010). These violations were institutional in character.

An interesting, and for us Norwegians depressing, example is a recent case in Norwegian cancer research, where it was revealed that several publications were based on extensive fabrication and falsification of data, published in the world's leading journals (see Michael, 2007). Articles were retracted, and an associate professor lost his PhD and his institutional positions. The most important thing to learn from this epistemological and ethical disaster has to do with institutions, internal controls, academic practices and routines. How was it possible to cover up this activity for six to seven years? The answer to that question obviously points to a lack of adequate institutional controls. The main culprit was, for instance, dependent on PhD supervision, research leaders, opponents to his dissertation, peer reviewers and 60 co-authors. The research had not been approved by independent regional committees for medical ethics, but research leaders had not checked that. In summary, this case is an example of system failure, pointing to the importance of several kinds of activities on different levels. It was investigated by external scientists and disclosed to the general public in a report. The report has been much used in Norwegian teaching and training for research integrity.

One important source for teaching of research ethics is to remember and learn from such negative cases. It encouraged several new initiatives in Norwegian research, and renewed several older ones. From 2007, the University of Oslo implemented a comprehensive action plan for the prevention of unethical practice in research and for stimulating good practice. It included a new handbook for good research practice, research ethics being incorporated into all PhD courses, and further empowered its committee to tackle problems in the field of research integrity.

The case also led to the development of a web-based pilot project ("IT initiative for sound scientific practice") made up of four elements: a scientific oath, documentation of new research projects, a declaration about authorship according to the Vancouver rules, and archiving of eletronic research material in a common university base. Last autumn, I finished an extensive documentation and evaluation of this project, having taken place in five of the University of Oslo's large departments (medicine, informatics, psychology, biology and history). I presented several recommendations with regard to teaching and training of research integrity on different levels, including adequate internal controls. I must be brief here, so let me just illustrate with one point. My data showed that less than 5% of the researchers had learnt about existing, relevant research ethical guidelines from a thesis supervisor. A challenge, therefore, is to train advisors, so that they also become advisors on ethics. It is strategic because the supervisor is so close to the actual research and can function as a role model for good scientific practice.

References

1. Boudon, R. (2001). *The Origin of Values.* New Brunswick (USA) and London: Transaction Publishers.
2. Habermas, J. (2003). *Truth and Justification.* Boston: MIT Press.
3. Harris, S. (1992). Japanese biological warfare research on humans. *Annals of the New York Academy of Science*, Vol. 666: 21–52.
4. Jones, J. (1993). *Bad Blood: The Tuskegee Syphilis Experiment.* NY: The Free Press.
5. Kalleberg, R. (2007). A reconstruction of the ethos of science. *Journal of Classical Sociology*, 7(2): 137–160.
6. Kalleberg, R. (2009). Can normative disputes be settled rationally? Cherkaoui, M., Hamilton, P. (eds.). *Raymond Boudon. A Life in Sociology* Vol. 2, pp. 251–270. Oxford, UK: The Bardwell Press.
7. Merton, R. (1973). *The Sociology of Science.* NY: The Free Press.
8. Michael (2007). Publication Series of The Norwegian Medical Society. (Research misconduct: lessons to be learned?) vol. 4: 7–61.
9. National Committees for Research Ethics in Norway. (www.etikkom.no). Accessed from http://www.etikkom.no/en/In-English/Publications/.
10. The National Commission for the Investigation of Scientific Misconduct. Accessed from http://www.etikkom.no/en/In-English/Scientific-Misconduct/FBIB/Ressurser/Granskingsrapporter/.
11. University of Oslo. (2008). Action Plan for the Prevention of Unethical Practice in Research. Dep. of Research Administration. (www.uio.no).
12. *The Lancet* (2010). Scientific fraud: action needed in China. Editorial, 9/1.
13. Walker, M. (1995). *Nazi Science.* NY: Plenum Press.
14. Wang, Q. (2006). Misconduct: China needs university ethics courses. *Nature*, 442, 132.

CHAPTER 33

DEVELOPING POLICIES FOR RCR TRAINING IN BRAZILIAN GRADUATE PROGRAMS: CURRENT CHALLENGES[1]

Sonia M R Vasconcelos

Compared to other Latin American countries, "Brazil is the region's giant in every sense of the word", and has "the most sophisticated and diversified science, technology and innovation system" (Foreign Affairs and International Trade Canada, 2008). The country has many projects in aviation, space and nuclear research, materials research, nanotechnology, biotechnology, biodiversity, biofuels, petroleum, and plays a strong role in the world's agricultural research sector, among other areas. As for matters related to the global economy, Brazil is a member of *Mercosul* (Southern Common Market), the G8+5 and the G20, and it has hundreds of international commercial partners. In terms of economic competitiveness, the growth observed for Brazil was considered the largest in the 2009 meeting of the World Economic Forum (Jorio *et al.*, 2010; Nature Editorial, 2010; Nature Materials, 2010).

In Latin America, Brazil has the highest proportion of its gross domestic product (GDP) invested in research and development (The World Bank,

---

[1]Thanks to all professors that have encouraged me to develop this RI project in Brazil, including Professor Hatisaburo Masuda (supervisor of the postdoctoral project), Martha Sorenson, Jacqueline Leta (IBqM/UFRJ); Adalberto Vieyra (IBCCF/UFRJ); and Marisa Palácios (IESC/UFRJ). Special thanks to Professor José Carlos Pinto (COPPE/UFRJ), who has offered great support for the RI project and helped with the focus group research at COPPE. I also thank Professor Nicholas Steneck (University of Michigan) for his collaboration and FAPERJ for the support and fellowship. I thank Professor Miguel Roig at St. John's University for his relevant comments on this contribution.

2007), which is now about 1.5%. In 2008, more than 50% of articles originating from Latin America and published in journals indexed in the Thomson Reuters database were from Brazil (Brazilian Ministry of Science and Technology, 2009). According to a recent interview with Sérgio Rezende, Brazil's Science and Technology Minister,

> So far, Brazilian scientists have mainly contributed to extending the frontiers of fundamental knowledge, and I believe this process will intensify further. Researchers are becoming more experienced, young people are being exposed to science of higher quality, and the infrastructure for research is improving. And in applied science and engineering, there will be an effort in all the areas that represent a priority at the international level. To list a few: biotechnology and nanotechnology; information and communication technologies; health; energy; agribusiness; biodiversity and natural resources; Amazonian and semi-arid regions; meteorology and climate change; space and nuclear research. (Nature Materials, 2010, p. 532)

As can be noted, Brazil today has an enormous potential for knowledge production and technology transfer in the global scene. In 2007, more than 80,000 PhDs were among those Brazilian researchers contributing to world knowledge and scientific research. Today, the accelerating rate at which PhD degrees are being awarded at Brazilian institutions, which will be consistent with that observed worldwide, points to the country's increasing international visibility and also to its potential for more international research collaborations (Anderson *et al.*, 2010). As expected, the increase in the number of Brazilian graduate students is associated with the growth in the number of publications in the major international scientific databases (De Meis, Arruda, and Guimarães, 2007).

However, when it comes to international science, there are important challenges to face. Among these is, the fact, that researchers from developed countries, especially those from an Anglophone tradition, appear to be better equipped to respond to the demands imposed by the "publish or perish" environment that has evolved in the areas of science and scholarship. First, the ability to write scientific papers in English favours the native English-speaking researchers in academia. As a consequence, getting published in English-language international journals for this group is less likely to involve the linguistic difficulties expected from non-native English-speaking authors, including those from Latin America.

In all of Latin America, including Brazil, the development of scientific writing skills in the language of science (i.e. English) is a critical issue that has not been adequately addressed for aspiring scientists. This is true if we assume that these authors, whose mother tongue is Spanish or Portuguese (in the case of Brazil) and for whom English is not even a second language, do not publish science in linguistically unfavourable settings and have little editorial support to write their manuscripts (Vasconcelos *et al.*, 2007). In Brazil, for example, data from 2005 showed that among the 22,900 Brazilian authors with publications in the Web of Science, only 51.4% claimed to have good writing skills in English. Interestingly, these researchers were also those with the higher number of publications, citations and h-indexes as compared to the group with less developed English writing skills (Vasconcelos *et al.*, 2008).

This is not just a Brazilian issue, as other publications have shown (Freeman and Robbins, 2006; Man *et al.*, 2004). The feeling of many scientists who are not native speakers of English who publish in English-only outlets is that they "don't compete on a level playing field when it comes to international science", and that "language and cultural barriers may be partly to blame" (Anon, 2002, p. 863). It is exactly in this publishing context that non-native English speaking scientists have to respond to demands for originality, not only of scientific data but also of text. The former is a concept internalised by authors in the scientific community, as manuscripts describing original research are the core of scientific knowledge. It is well known that this feature represents one of the major factors in accepting a submission. Peer acknowledgment of data originality, from an objective perspective, would be more of a scientific than a cultural issue — though one cannot assume this evaluation process is free of cultural bias. The evaluation of originality of a scientific text, on the other hand, does not seem to have achieved nearly the same degree of international consensus and, in fact, has become an increasingly sensitive issue among peers. This may be a reasonable assumption, given that textual originality in today's science involves Anglophone notions of plagiarism, which are those that define current editorial policies.

In Brazil, focus group research data, although qualitative in nature, have indicated that internalising the concept of plagiarism in science cannot be taken for granted. A recent comparison of views of plagiarism between senior researchers ($n = 16$) and graduate students ($n = 17$) who participated in focus groups conducted in 2008 (Vasconcelos *et al.*, 2009) and 2010 (unpublished ongoing research) are quite similar. The classic definition shared by Anglophone countries focusing on inappropriate use of others' ideas, results and words did not show itself to be clear cut among respondents. When asked about the definition of plagiarism, most were doubtful about the inclusion of words. They attributed different weights to plagiarism of data and ideas as compared to plagiarism of words in science.

In fact, when it comes to publications, addressing textual plagiarism in science raises not only ethical challenges but also cross-cultural challenges. Among other things, this means that authors from non-Anglophone countries should be well informed about the constraints imposed by writing up research in this context. Yet, it does not seem that the scientific community at large has achieved the consensus required to avoid misunderstandings among researchers. The following comment by Turkish researchers accused of plagiarism reflect part of the problem:

> Borrowing sentences in the part of a paper that simply helps to better introduce the problem should not be seen as plagiarism. Even if our introductions are not entirely original, our results are — and these are the most important part of any scientific paper. In the current climate of "publish or perish", we are under pressure to publish our findings... (Yilmaz, 2007).

In Brazil, a large national survey would be necessary in order to obtain a broad picture of the relationship between the various relevant variables, such as cultural assumptions, linguistic barriers, and notions of plagiarism in science. Even so, the fact that plagiarism is a worldwide issue is indisputable and it has emerged as a significant concern in the research community. Today, this practice has increasingly found itself among the reasons for retractions in science (Corbyn, 2009). In countries where research integrity (RI) initiatives have been part of the agenda of universities and funding agencies, attempts to address and detect plagiarism are part of the approach. It is thus expected that junior researchers from these countries — including the USA, England, Canada, Australia, Singapore, and many others — have been better prepared to respond to current demands for doing research in today's "publish or perish" environment.

In Latin America and in other regions that have had little engagement in RI discussions, the panorama is certainly different. In Brazil, for example, apart from the challenges related to publishing science in a linguistically unfavourable environment, addressing plagiarism within the context of RI is another challenge. In fact, RI has received scant attention in the country. Formal policies are not clearly established, and Brazilian researchers, educators, and policy makers have not engaged in the recent RI conversations led by the USA, European and other developed countries. Among the challenges to be met in this context are the following:

- Too few professors involved in RI and related initiatives to start these discussions in Brazilian institutions.
- Developing RCR training in graduate programs is not required and, so far, does not seem to be a concern.
- Attention to formal training in research ethics at large is still scant. Even bioethics training is not mandatory in most graduate programs in the biomedical sciences.
- The definition of misconduct established by most international organisations is not widely publicised in Brazilian research institutions, and junior researchers in graduate programs have little exposure to current dialogues on questionable research practices.
- As in many other countries, the current discussion and publication policies on authorship issues need to be addressed among graduate students, as institutional/departmental cultures establish different practices.

However, the increasing attention to quality in the evaluation of scientific research in Brazil should play a major part in the insertion of the country into the current debate on RI and Responsible Conduct of Research (RCR). In the recently published *Nature* news article "High Hopes for Brazilian Science" (Petherick, 2010), we read that "the chances are good that scientists will get much of what they ask for on their consensus wish list" (p. 675). I believe this wish list will include more opportunities to reduce cultural, linguistic and ethical gaps in the context of Brazilian science, and the following initiatives may be evidence that efforts have been and will continue to be made to fill up those gaps:

- The Carlos Chagas Filho Foundation for Research Support in the State of Rio de Janeiro (FAPERJ) has recently approved a 5-year fellowship for a postdoctoral research project focusing on developing RCR policies for graduate programs in the sciences through a pilot project at one of the biggest federal universities in the country.
- A distance learning course on research ethics, which includes a module on RI, has been developed by a group of Brazilian researchers (mostly working on bioethics), which has been coordinated by The National School of Public Health of the Oswaldo Cruz Foundation (ENSP/FIOCRUZ), a major research center in Brazil.

- The Brazilian Society for Biochemistry and Molecular Biology (SBBq) included a symposium entitled "Research Integrity: Is Plagiarism a Problem?" in its 39th Annual Meeting. The symposium encouraged discussion in the research groups of professors attending the meeting.
- After the end of the SBBq Meeting, the Society included recommendations for Brazilian graduate programs to include RI initiatives in their agenda. These recommendations are part of SBBq's suggestions for the upcoming Brazilian National Graduate Program Plan (2011–2020).

Finally, RI and RCR initiatives in Brazil are still in their early stages, but the country has the greatest potential to create the proper climate for these initiatives not only within its own borders, but also throughout Latin America. Brazil thus should be poised to play a more active role in international dialogues on research ethics and integrity.

References

1. Anderson, M. S., Chiteng, Kot, F., Jie, Y., Kamata, T., Kuzhabekova, A., Lepkowski, C. C. , Shaw, M. A., Sorenson, M. M. and Vasconcelos, S. M. R. (2010). Differences in national approaches to doctoral education: Implications for international research collaborations. In Anderson, Melissa, S., and Steneck, Nicholas, H. (eds.), *International Research Collaborations: Much to be Gained, Many Ways to Get in Trouble*, New York: Routledge.
2. Anon (2002). Breaking down the barriers. *Nature*, 419: 863.
3. Brazilian Ministry of Science and Technology. Indicators coordination (2009). http://www.mct.gov.br/index.php/content/view/5710.html.
4. Corbyn, Z. (2009). Retractions up tenfold. *Times Higher Education.* http://www.timeshighereducation.co.uk/story.asp?storycode=407838.
5. De Meis, L., Arruda, A. P. and Guimarães, J. A. (2007). The impact of science in Brazil. *IUBMB Life*, 59: 227–234.
6. Foreign Affairs and International Trade Canada (2008). *Brazil: A Global Commerce Strategy Priority Market.* http://www.international.gc.ca/commerce/strategy-strategie/r5.aspx.
7. Freeman, P, and Robbins, A. (2006). The publishing gap between rich and poor: the focus of AuthorAID. *J Pub Heal Pol*, 27: 196–203.
8. Jorio, A., Barreto, F. C. S., Sampaio, J. F. and Chacham, H. (2010). Brazilian science towards a phase transition, *Nature Materials*, 9.

9. Man, J. P. *et al.* (2004). Why do some countries publish more than others? An international comparison of research funding, English proficiency and publication output in highly ranked general medical journals. *Eur J Epidemiol*, 19: 811–817.
10. Nature Editorial (2010). Brazil's biotech boom. *Nature*, 466: 295.
11. Petherick, A. (2010). High hopes for Brazilian science. *Nature*, 465: 674–675.
12. Vasconcelos, S. M. R., Leta, J. and Sorenson, M. M. (2007). Scientist-friendly policies for non-native English-speaking authors: timely and welcome. *Braz J Med Biol Res*, 40: 743–747.
13. Vasconcelos, S. M. R., Sorenson, M. M., Leta, J., Batista, P. D. and Sant'Ana, M. (2008). *Researchers' writing competence: A bottleneck in the publication of Latin-American science? EMBO Rep*, 9: 700–702.
14. Vasconcelos, S. M. R., Leta, J., Costa, L., Pinto, A. and Sorenson, M. M. (2009). Discussing plagiarism in Latin American science: Brazilian researchers begin to address an ethical issue. *EMBO Reports*, 10: 677–682.
15. World Bank (2007). *Global Forum: Building Science, Technology, and Innovation Capacity for Sustainable Growth and Poverty Reduction.* Washington, DC, USA: The World Bank.
16. Yilmaz, I. (2007). Plagiarism? No, we're just borrowing better English. Correspondence, *Nature*, 449: 658.

SECTION VI

INDIVIDUAL APPROACHES TO FOSTERING INTEGRITY IN RESEARCH

INTRODUCTION

Given the many options available for fostering integrity in research, there is probably no single best approach. Fostering integrity is an individual responsibility, and the way individuals go about this important task depends on personal styles and goals. The papers in this section provide a number of examples of the ways researcher-teachers in different countries and settings have taken on the challenge of imparting the understandings and skills needed to be responsible researchers.

Based on wide personal experience, Vasgird suggests ways in which online tutorials, cases and videos can be used to enrich and enliven instruction. McKellar provides other examples of how case studies are used in training researchers in Australia. Roland takes a different approach to teaching, focusing on core competencies, such as critical thinking, vision, problem-solving and awareness of quality.

Three papers from Japan illustrate how different individuals face the challenges of reaching out to students and researchers in one country. Iseda describes how he uses the sociology of research to help Japanese students understand the concept of "professional integrity", a concept that is difficult to translate into Japanese. Ichikawa and Motojima describe their use of the US-developed CITI programme to meet Japan's mandate for ethics education for anyone who does human subjects research, not only translating the text but changing the content to reflect local Japanese requirements. Asashima describes yet another effort underway in Japan — the publication of special books designed to encourage young researchers to adopt high standards for integrity in their work.

There are also significant time and audience variations in responsible research courses. Axelsen provides the *why*, *how*, *what*, *does it work*, and *was it needed* for a special course on Good Scientific Practice developed at Denmark's Statens Serum Institute. The course targets a small audience (8–10 students) and is taught in five three-hour sessions. Bain describes an effort to train many students at the Australian National University, where two-hour workshops have been designed in a way that allow them to be offered by different instructors. Ritter and Webster explain the origin, rational and methods of for-credit coursework on integrity at Imperial College in the UK. All three of these presentations mention the role of national codes or standards in course development.

Based on these examples, the general observations set out in Section 5, conference discussions and a special post-conference workshop, delegates reached significant agreement on the challenges and steps that need to be taken to address them, and summarised these in the Workshop Report at the end of this Section.

CHAPTER 34

ONLINE RCR TRAINING AND THE USE OF CASE STUDY VIDEOS

Daniel R Vasgird

Integrity and responsibility are words with profound implications, especially for those who participate in the global community of science. They cut across time and culture, and yet their fruition in the guise of an ethical lifestyle depends on one's ability to conceive an ideal, aspire to it, and abide by its principles to the best of one's ability. In turn, the truly ethical society depends on its ability to crystallise that conception of the ideal in the hearts and minds of its practitioners. As Richard Livingstone once said, "One is apt to think of moral failure as due to weakness of character; more often it is due to an inadequate ideal." It is the role of societies (with science being one) that wish to flourish to provide the means for their constituents to truly internalise their highest and most worthy ideals.

Research, representing the systematic side of science, generally flourishes when the public that supports it and ultimately makes use of its products has high regard for its ways and means. Science and the global public can be seen in a contractual relationship, sometimes explicit and sometimes implicit. Therefore, every effort must be made to bolster the invaluable commodities of respect and trust. Many scientific organisations have internalised this insight and expanded their research integrity initiatives concomitantly to meet the needs of a more expansive, demanding and competitive era. The nurturing of research integrity for these entities has become a forthright rather than presumed endeavour. With this interest in and concern about research integrity, a wide array of approaches to the enhancement of research responsibility has evolved. This portion of the 2nd World Conference on Research Integrity "Training for Responsible Research" track reviewed two educational techniques that are being widely used to enhance responsible conduct of research (RCR) education and training programs at many institutions: online training and the use of case study videos.

I personally have two challenges in mind when I develop an RCR education program. There are always variations based on the audience and context, but generally I focus on stirring an informed empathic response in as succinct a fashion as possible. What I have found with succintness and empathy as my guide is that both online tutorials and case study videos have immense value. In time-pressed environments (are any not these days?), I have made use of them with great effectiveness.

Online tutorials can be accessed through the Web and are an excellent way to survey the terminology and concepts which provide the vocabulary and context for interactive RCR involvement in a course environment (classroom or online) or in group meetings, journal clubs or even one-on-one mentoring situations. They are usually developed in module sets based on RCR core areas such as authorship, mentoring, conflict of interest, research misconduct, etc. (See the ORI education resources website for a listing of many such tutorials: http://ori.dhhs.gov/education/products/).

Most experts in RCR education agree that online tutorials should be thought of as a tandem resource to supplement real-time learning environments that allow for spontaneous give-and-take responses and discourse between participants. Their true utility is in bringing participants up-to-speed on terminology and concepts for better use of valuable live interaction time. I generally assign an RCR online topic-specific tutorial for review a week or so prior to interactive involvement, whether for a class discussion or for a lecture with follow-up question and answer time. Lectures and discussion sessions are normally focused on very specific case studies and/or topics, the presumption being that the tutorial review has provided the conceptual foundation so that we can jump right into the core of the matter at hand.

RCR online tutorials by and large are organised in five parts: introduction, case study section, foundation text, resources and conclusion. The *introduction* will lay out the objectives for the tutorial, explain to users what to expect, sometimes give a brief video introduction, and lay out a series of optional challenge questions. Each module will next usually have a central *case or cases* made up of a narrative of events. The events of the cases are realistic fabrications with questions and answers related to best practices. The *foundation text* will include all relevant background text material for the topic including, but not limited to, definitions, examples, policy statements, perhaps videos of other expert opinions and experience, and practical tools. It will contain the basic pieces of information that are necessary for the users to read to have a fundamental understanding of the topic. Techniques include the use of side-bar material, and the option to jump around the text via a linked table of contents. A *resources section* will provide a list of policy statements and readings related to the topic. The *conclusion* will offer a synopsis of the tutorial and frequently review the objectives.

I often approach the challenge of empathic response in a two-step fashion: insight and case study videos. I begin by drawing on a bank of quotations compiled over the years for the singular purpose of inducing individuals to think outside themselves. Keep in mind that ultimately what the RCR educator is aiming at is developing a free-standing culture of social responsibility where people will think about the consequences of behaviour prior to action, and hopefully in relation to the broader community around them. The empathic response I am looking for is really a set-up for the rest of my presentation to stir the individual to think from other points of view. A couple of examples are included here with a comment for each, one from Albert Schweitzer, the Nobel Peace Prize winner, and the other from Randy Cohen who wrote The Ethicist column for *The New York Times*:

> Schweitzer: "The first step in the evolution of ethics is a sense of solidarity with other human beings." Oftentimes these words will lead to a thought in the audience that it is hard to think about consequences for others unless one feels for them one does for the same as in family.
> Cohen: "Ethics primarily concerns the effects of our actions on others." This will often conjure a discussion about the importance of thinking seriously about the consequences of actions and their ripple effects.

After setting the stage, one can move forward on the empathic response path by utilising a recent and evolving tool, case study videos. Case studies have long been recognised as an effective tool for generating discussion and insight regarding ethical decision-making. The value of case study presentation and analysis is of course that they are narratives based on realistic portrayals, prompting individuals to project themselves into situations that they can identify with. There is a range of such videos available these days, many of which can be found by going to the previously referred-to ORI Educational Resources site (http://ori.dhhs.gov/education/products/).

One type of these case study videos has come to be known as "trigger videos" and can be interjected early on in an RCR presentation. They are brief (~3 minutes) and usually open-ended in terms of resolution. Considering though that RCR education sessions often face time constraints due to the nature of the academic research environment, they are a quick and efficient tool for stimulating discussion and demonstrating how research practitioners can unintentionally find themselves in ethically problematic situations. A sampling of "trigger videos" can be found at my institution's website at: http://oric.research.wvu.edu/rcr_train/rcr_videos.

In sum, over the last few decades scientific associations, governments and academic institutions have become increasingly more concerned and proactive in advocating the responsible conduct of research, in light of the increasing prominence and growing resources of global science. It is crucial to understand that the global public has placed a tremendous responsibility in the hands of science — its hope for the future — but it has expectations in return. At the top of that list is a sense of responsibility from which flows that oft-referred to sense of trust. Responsible conduct of research education initiatives are aimed at developing and bolstering that culture of responsibility in science. Some of the tools to pursue this have been described here for use as appropriate and needed.

CHAPTER 35

USE OF CASE STUDIES IN TRAINING STUDENTS AND PRACTITIONERS IN RESPONSIBLE RESEARCH PRACTICE

Bruce H J McKellar

I have used case studies to raise awareness of the following groups of staff about ethics and research integrity:

- my peers in the School of Physics, University of Melbourne;
- graduate students in Physics at the University of Melbourne; and
- mid-career staff in the Australian Defence Science and Technology Organisation, including scientists, engineers, and administrative staff.

The courses have ranged from 2 to 4 hours total duration.

I begin by motivating the course, giving examples of unethical research practice, preferably ones that have recently been in the news. I have found it useful to give a brief account of the different approaches to ethics, because I want to start the audience thinking about their own approach to ethics, and I encourage that by emphasising that even when ethical behaviour is prescribed, for example, in a code of conduct, you do need to establish your personal position on it. This is a preliminary to the discussion of case studies, assigning one or two to each small group. For each case, the group then reports their discussion in a short presentation. There is also a window of opportunity for others to provide new insights into the case. At the conclusion, there is an opportunity for me to summarise and draw out some concluding points.

One case study illustrates conflict of interest, which I drew from the American Physical Society case studies.[1]

> A student finishes a PhD working on a problem that has aspects that are directly patentable and that can solve a major problem in the disk drive industry. He or she arrives at the new job with a major disk drive manufacturer, and discovers that the work done as a student, which is in the midst of the patent process at the university, will solve a problem at his new company. If he reveals what he knows to his new employer he will be an immediate hero, but will compromise the patent process at his original institution. His actions could have important financial implications for the original institution, and perhaps for the student.

What should the student do?

Another study illustrates questions that arise in a case of falsification. This case is drawn from my own experience.

> You are a PhD supervisor who gets a letter from a colleague, who pointed out that the theoretical curve in a paper published by you and one of your student was different from his own calculations, but it was in good agreement with the experimental data. The student now has a junior faculty position at another institution.

What do you do?

Yet another case study, drawn from the news, illustrates many facets of research misconduct.

You are the President of a University. A researcher in your University is accused of

- Fabricating data
- Using a grant to support research unrelated to the grant proposal
- Publishing a paper drafted by a student as his sole work
- Adding his wife, who was not involved in the research, to the list of authors on one of the papers to enhance her prospects of getting a grant
- Sacking a postdoctorate who complained about the above

What do you do?

As my use of case studies evolved, I found it important that the group discussion of the cases be moderated. This increases the effectiveness because it ensures that all participants have an opportunity to give

[1] http://www.aps.org/programs/education/ethics/interest/confidentiality2.cfm.

their views. But it also allows the possibility of giving the moderator some questions to use to stimulate the discussion. For example, for the conflict of interest case, one of the questions suggested was, "What are the relevant ethical considerations in this dilemma?" The answer could depend on missing information, such as the nature of the contract signed by the student with his institution, if any, and the nature of its intellectual property rules. This information may have an impact on what the correct ethical response should be.

I have found that the use of case studies works best when the discussion groups are small. Used as the basis for discussions, case studies are effective in making the students think about their personal ethical attitude, and how they may react in particular situations.

My experience is that it is important to discuss the philosophical basis of ethics and challenge students to articulate their personal ethical position and how they get to it. Then the use of case studies puts teaching about ethics and responsible conduct of research into a concrete situation, so that some thinking and discussion is required of the students.

Perhaps it is because of the way I have approached my teaching of research ethics and responsible conduct of research, but for me, it is now impossible to imagine teaching such a course without relying on case studies.

CHAPTER 36

RÉFLEXIVES® INTEGRATED TRAINING PROGRAM FOR PhD STUDENTS AND THEIR SUPERVISORS: QUALITY, INTEGRITY AND RESPONSIBLE CONDUCT OF RESEARCH

Marie-Claude Roland

From the beginning (1995) our hypothesis has been that training through research can provide doctoral students with core competencies expected from young professionals and autonomous and responsible citizens: critical thinking, vision, problem-solving, awareness of quality, integrity. A 2004 EU report under the direction of Bourgeois *et al.* confirmed this hypothesis was still relevant, under strict pedagogical principles.

1 BACKGROUND PRINCIPLES

Loosely defined stages are common to most research projects. Most authors argue that it is incumbent upon the supervisor to bridge the gaps in communication during the various stages of research by requesting regular meetings and updates, and to develop a basic level of collegiality to make the research project successful. The roles expected and assumed by students and supervisors — as guide, project manager, or "critical friend" — structure the relationship and the strategies for supervision.

We defend the following ideas:

- The research project plays a fundamental role in structuring thinking, knowledge acquisition and elaboration; it strongly contributes to the development of competencies like autonomy, critical analysis, integrity and responsibility and the capacity to synthesise ideas and experience.

- The methodological approach to structuring a research project is key to a successful relationship between supervisors and students.
- Epistemology and ethics are inseparable from research practice, and therefore education in the Responsible Conduct of Research is the responsibility of researchers and should be addressed internally, by researchers themselves in the research environment/institution and not by external consultants.

2 OBJECTIVES

From the supervisor's idea to the PhD student's research project — opening transaction spaces to foster relations of cooperation and effective communication, to reflect on research practices and on the Responsible Conduct of Research:

1. Training young professionals to

- "Have the big picture" — capacity to situate their project and to situate themselves;
- Conduct their research in a responsible manner, behave with integrity and implement quality criteria;
- Develop critical thinking and analysis, and become effective readers and reviewers;
- Communicate effectively with their peers and with society; and
- Build an Individual Development Plan and a Portfolio.

2. Training supervisors to

- Think critically about their role as supervisors, mentors, and role models;
- Ensure that the PhD student will take ownership of his/her research and professional projects; and
- Formulate their competencies.

3 STRATEGY

To work within the scientific community (institutions, labs) and involve supervisors in the training process in order to:

- Open spaces of transaction to build the PhD student's research project;

- Encourage dialogue between disciplines and focusing on scientific and societal issues;
- Assess the gap between norms and practices in the conduct of research;
- Develop reflective practice to analyse research practices, professional practices and identify the competencies involved; and
- Foster mutual learning and learning in action.

4 METHODS AND TOOLS

- Each workshop includes 4 supervisor/PhD student pairs and is facilitated by 2 researchers who are not involved in the projects but who are trained in communication and mediation techniques.
- Theoretical presentations and real-work situations (work on students' research projects).
- Mind-mapping to foster creativity, linguistics, critical reading and analysis.
- Reflective practice to encourage reflection on and in action — participants are encouraged to use a journal/learning log to record their thoughts and feelings and work on problems.

5 DESCRIPTION OF THE PROGRAM

The training program is conducted by the Réflexives® Community of Practice involving researchers from various disciplines (mathematics, life sciences and social sciences) whose job is to "integrate, innovate and facilitate". It is part of an intervention-research project which involves researchers as participants and generates innovations, thus ensuring that the program is relevant, adaptable and applicable to practice.

The Réflexives® Training Program meets the recommendations of the report "Integrity in Scientific Research: Creating an Environment That Promotes Responsible Conduct" produced by the Committee on Assessing Integrity in Research Environments, National Research Council, Institute of Medicine, the National Academies Press at: http://www.nap.edu/catalog/10430.html (Chapter 5 — Promoting Integrity in Research through Education, pp. 85–86)

The training program comprises a 5-day core seminar and complementary technical workshops.

5.1 *The Core Seminar: (5 Days — at a Specific Venue to Encourage Interactions)*

During the week, supervisors and PhD students participate in 4 different types of sessions which are all focused on the conduct of research, on practices and on the values — personal and professional — involved (Figure 1.). The different sessions all share the same approaches: dialogue, reflection on the gap between norms and practice, analysis of written and oral discourse, and skill development:

- *Project building*: workshops aim at fostering creativity, autonomy, quality and integrity, and effective communication between supervisor and student.

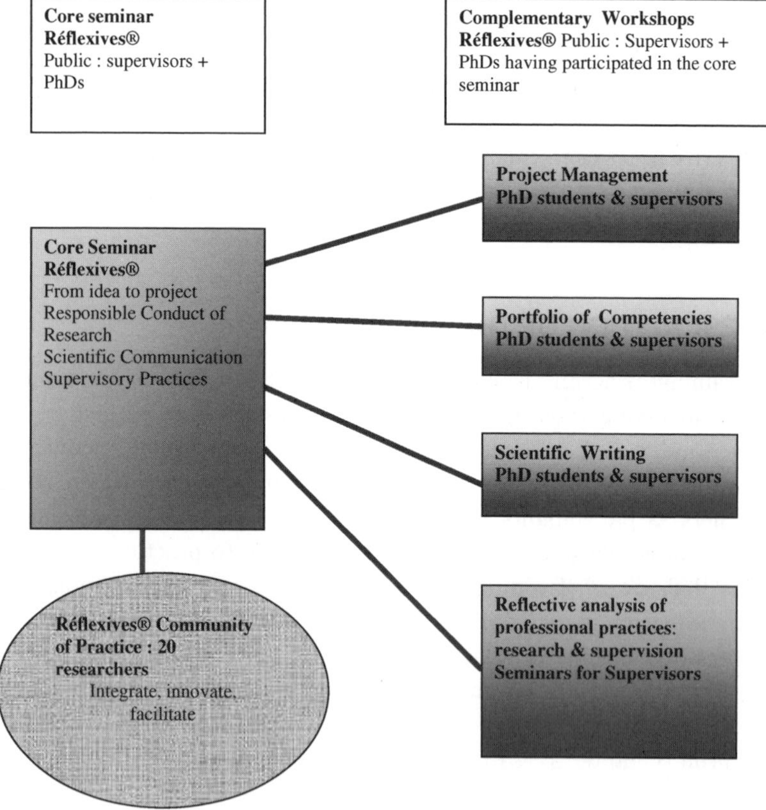

Figure 1. The Réflexives® Integrated Training Program at a glance.

- *Science communication*: the participants' own written productions are critically analysed by the group, theoretical presentations alternate with open discussions on practices, participants are encouraged to reflect on the ethical dimensions of their work and on the codes and guidelines that govern their field of research and to provide rationales for their choice of action. Their responsibility as authors is addressed.
- *Development of competencies*: theoretical presentations, individual and collective analysis, reflective practice are used to help participants to identify their competencies and initiate the development of a portfolio.
- *Reflective practice*: every day, participants have a specific session at the end of the day to reflect on the day's events, on scientific practices as well as social interactions within the group and to develop ethical reasoning and sensitivity.

5.2 *The Complementary Workshops (2 Days Each, in the Year Following the Core Seminar)*

- Both supervisors and PhD students attend workshops on project management, scientific writing, and building a portfolio.
- Supervisors are encouraged to attend workshops on Reflective Analysis of Professional Practice at regular intervals, to foster sustainability of the training.

CHAPTER 37

HOW TO TEACH RESEARCH INTEGRITY WITHOUT THE NOTION: ATTEMPTS IN JAPAN

Tetsuji Iseda

In this paper, I explore the problem caused by a cultural difference in terms of research integrity education. Present-day science is a global endeavour, but each scientist is raised in his/her own cultural context, and a successful research integrity education should be tied to such a context; otherwise, the rules would not make sense. Thus, if we want to create a truly global scheme of research integrity, we need to be sensitive to cultural differences. This paper is an attempt to contribute to the discussion toward such a scheme. Part of the paper reports actual educational practice, and part of it presents thoughts on a possible educational scheme.

I speak from my experiences in Japan. I have taught research ethics classes in Nagoya University for several years. It is a class called "science and engineering ethics" and is taught by four teachers (two philosophers, one engineer and one primatologist). The students are sophomore students from various fields including the humanities, social sciences, science and engineering.

My role as a philosopher who specialised in both philosophy of science and ethics was to give a theoretical background for research ethics. Around that time (2006 or so), the notion of research integrity was introduced to Japan, and most universities in Japan, including Nagoya University, established their own guidelines for research integrity. I served as a committee member for establishing the guidelines, and tried to explain the notion of research integrity in my classes.

What I found was that conveying the notion of "research integrity" to Japanese students (and other researchers, for that matter) had a particular problem. The very notion of "research integrity" was absent until quite recently. The most common expression used to translate "research integrity" is the newly-created term *kenkyuu kousei* (研究公正), which literally means "fair treatment in research". The word "integrity" in this context did not have exact corresponding Japanese terms, but "professional integrity" is often emphasised in many other professional ethics fields, which leads to similar translational problems occurs. However, the notion of integrity is not as central in other fields as is the case with research integrity.

Is there any problem with teaching a research ethics class without the notion of integrity? Integrity represents the positive motivational structure toward research ethics. Without it, ethics education can be easily reduced to avoidance of FFP (fabrication, falsification and plagiarism) and similar clearly definable misconducts. I do not think such an impoverished view of ethics promotes ethical conduct in research in general.

One attempt made in my classes at motivating students without the notion of integrity is to give a short lecture on the sociology of scientific research. Scientific research as a communal activity is based on particular reward systems like priority and credit. FFP takes advantage of these systems, and as such threatens the normal functioning of the scientific community. FFP also destroys the trusting relationship between science as a profession and the larger society, on which all scientific activities rely. This amounts to incorporating an element of professional ethics into research ethics education.

The sociological background was partly successful in making sense of FFP and other misconducts, but did not seem to be satisfactory in terms of positively motivating students. Engineering ethics education in Japan faced similar problems when fellow teachers and I tried to set up a positive image of engineering responsibility. Looking for something close to integrity, we come up with the notion of *hokori* (誇り) which means "pride" or "self-esteem" in Japanese (Iseda 2008). Having pride is of course different from having integrity, but they share a positive attitude toward ethics and motivate people to set higher standards for themselves.

This way of motivating students works for engineering students, and possibly for students in other fields too. There are many ways in which researchers can be proud of themselves: as a member of the scientific community in which great scientists everyone knows take part, or as an expert in producing original reliable knowledge which society is willing to fund. Such positive emotions will naturally lead researchers to protect and promote the source of their pride. I have not tried the "pride" education on

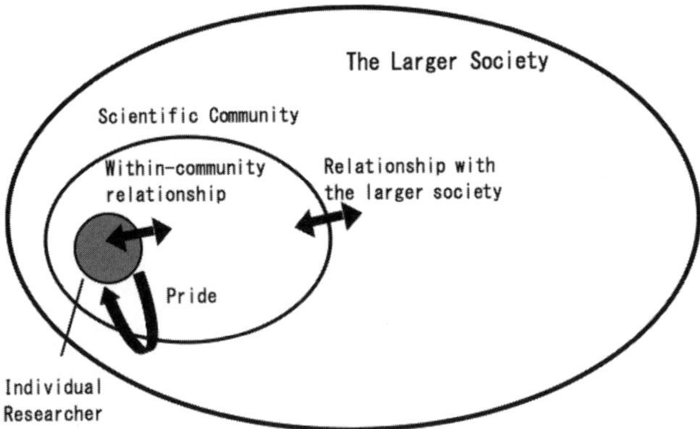

Fig. 1. The motivational structure of pride-based research ethics.

general science major students, but personal conversations seem to show that there is potential.

The motivational structure that puts together two sociological considerations and the pride-based consideration can be schematised as follows:

The largest motivation for an individual researcher to engage in responsible conduct comes from her relationship with the scientific community and his/her understanding of the way it works. Another motivation for responsible conduct comes from his/her understanding of the relationship between the scientific community and the larger society. The motivation (or at least part of it) to set the highest standard of responsible conduct comes from his/her self-regarding motivation like pride or self-esteem (Figure 1.).

These observations and proposals have an implication for the attempt to establish a global standard of research integrity education. If my observation is right, to require the understanding of the notion of research will place an unreasonable burden on a student if she is from a culture without the notion of "integrity". An alternative requirement, suggested by my proposal, is to have a motivational structure that makes sense within the culture and that has similar behavioural effects as understanding the notion of integrity. This will lead to a truly culturally sensitive research integrity scheme.

References

1. Iseda, T. (2008). How should we foster the professional integrity of engineers in Japan? A pride-based approach. *Science and Engineering Ethics*, 14(2): 165–176.

CHAPTER 38

CREATING THE CITI-JAPAN PROGRAM FOR WEB-BASED TRAINING: WHERE ETHICS, LAW AND SCIENCE EXPERTS MEET[1]

Iekuni Ichikawa and Masaru Motojima

When compared to the USA, no other country in the world possesses the critical mass of experts necessary to create quality Web-based teaching materials (from here on called "modules"), like those produced by the CITI (Collaborative Institutional Training Initiative) program. The number of medical schools in Japan, for example, is half of the number in the USA in proportion to the population, and the number of faculty members per school is about one-fifth. Moreover, many will be surprised to know that in Japanese society, where its citizenry carries Western and Asian cultures in variable combinations, there is no established standard of medical ethics to teach. In general, this island nation also does not traditionally boast strong initiative from the ground level, so that even when its national government mandated its researchers who work on human subjects to take a training course for research ethics, the private sector has been extremely slow to follow through. One typical example of this can be seen in the humanitarian response to the Indian Ocean earthquake in 2004 (Table 1).

[1]Our success thus far owes immensely to the generosity of the CITI Program, which permitted the use of their modules in our efforts to create our own modules and to provide us with the software and hardware to run our program. We are also in debt to the CITI Program support staff, who have replied to our multiple requests promptly and gracefully. Presentation at the 2nd World Conference on Research Integrity was supported by a grant from the Japan Society for the Promotion of Science. The Authors also want to express their sincere appreciation to Akiko Ichikawa for her editorial assistance.

Table 1. Among of monetary aid given to victims of the 2004 Indian Ocean earthquake.

	USA	Germany	Japan
		Million, USD	
Government	277	82	539
Private	1,875	580	34

Data from references 1–3 are combined.

Whereas the Japanese government donated twice and six times the amount given by the US and German governments, respectively, to the afflicted countries, the Japanese private sector donated only one-sixtieth and one-twentieth of those of the US and German private sectors. The amount from the US private sector was nearly 7 times more than that from the US government, whereas the amount from the Japanese private sector was only one-sixteenth of that from its national government.

With regard to the lack of initiative in the Japanese private sector, an expert panel of the Science Council of Japan made an interesting comment in their recommendation in handling misconduct in research (Figure 1)[4] in year 2005. It stated that it "recommends not to directly apply Euro-American rules to Japanese researchers as they do not have autonomy comparable to Euro-American researchers..." Japanese researchers lack autonomy. Then, the logic goes, there is no reason to have an oversight body to impose penalties on misconduct. Thus, there are never court rulings and no official records, only news articles, which are often claimed to be inaccurate. Thus, there is no officially accepted history of precedents that can be used as teaching material, which, in turn, deprives the field the chance to build a basis of autonomy, thereby creating a vicious cycle.

Under these circumstances, the Japanese government made an announcement in 2008, mandating researchers working on human subjects to take an ethics education.[5] Many private institutions have simply taken this announcement as an instruction to give their researchers opportunities to learn government regulations and guidelines, rather than ethics, taking a rather passive "let-the-government-instruct-us-and-we-will-follow-it" attitude.

Although Japan does not have sufficient numbers of experts volunteering their time to create teaching modules in the same way that the USA has, the country still needs modules that are equivalent in quality to those

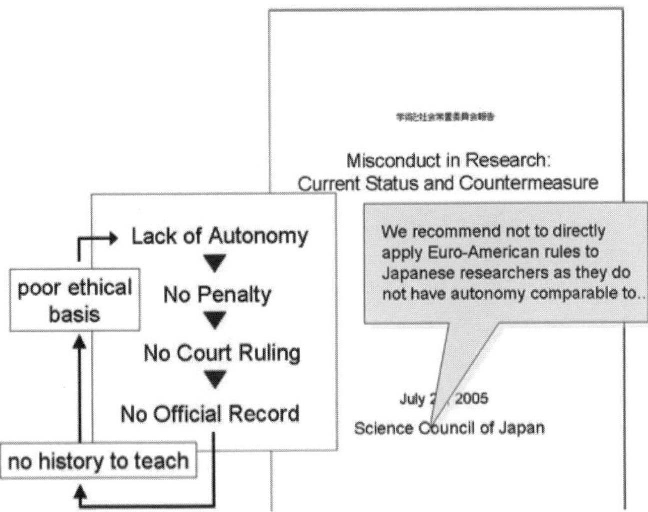

Figure 1. The report to the Japanese government prepared by an expert panel organised by the Science Council of Japan on misconduct in research. The document analyses the current status in Japan and makes recommendation for improvement.

used by US researchers. To achieve this goal and create high quality Web-based teaching modules with limited support from experts, we took to the following approach (Figure 2). Our scheme was initially presented by the first author as discussion material at the week-long faculty development seminar, "Program: Leaders in Healthcare Education," offered by Harvard Macy Institute in Boston[6] in 2005. Thus, we first imported CITI's teaching modules that were already on the Internet to translate them into Japanese. These are not only authored by world experts, but reviewed by multiple referees, and even more importantly revised in accordance with many comments from actual users. The modules also provided good guidance for what content, to whom, the level of typical learners, to what extent, how to teach, and what makes for user-friendliness. Once translated, the modules were "Japanified", or adopted to reflect Japanese regulations, culture, and history. The resulting modules were then submitted to Japanese experts in the field, albeit scarce in number, for review. The Japanified modules thus adjusted were then reviewed by bilingual individuals designated by CITI who were familiar with the field to verify that those Japanified modules still maintained the direction set forth in the original version. This step is important for Japanese learners as the ethical standard they are being

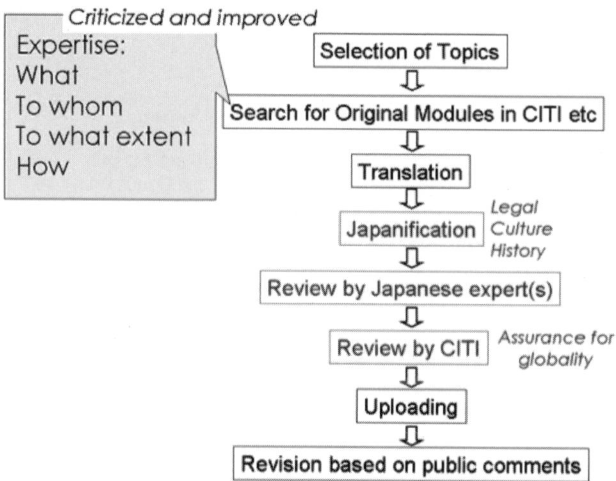

Figure 2. Outline of the process of generating Japanese versions of web-based research ethics education modules. The process involves (1) use of CITI modules as their bases, which have already been through reviews and revisions, (2) translation into Japanese, (3) incorporation of Japanese laws, regulations, cultures and history, as necessary, (4) revisions based on the comments from Japanese experts in the field, (5) review and assurance by CITI for their compatibility to global standards, (6) Internet uploading, and (7) revisions based on comments from users.

taught to commit to will be one that is globally accepted. The modules were then uploaded onto the Internet. We are currently at the stage of revising the first edition.

Our home page, www.CITIprogram.jp, (Figure 3) is accessible from the CITI home page (Figure 4). Thus far, experts from more than 13 institutions have participated in creating Japanese modules. The number of institutions using these modules is also increasing, which includes not only medical schools and research institutions, but also professional societies (Figure 4). Although the pace of overall usage after 18 months of being open to the public is still slow, as was the case for CITI itself initially when it started in year 2000, it is reflective of steady growth. We speculate that our success thus far can be attributed to the following, some of which we believe also apply to the success of the CITI Program:

1. Including the authors of the original US modules, experts from both liberal arts and science fields participated in the generation of modules, so that they could be made acceptable to Japanese researchers.

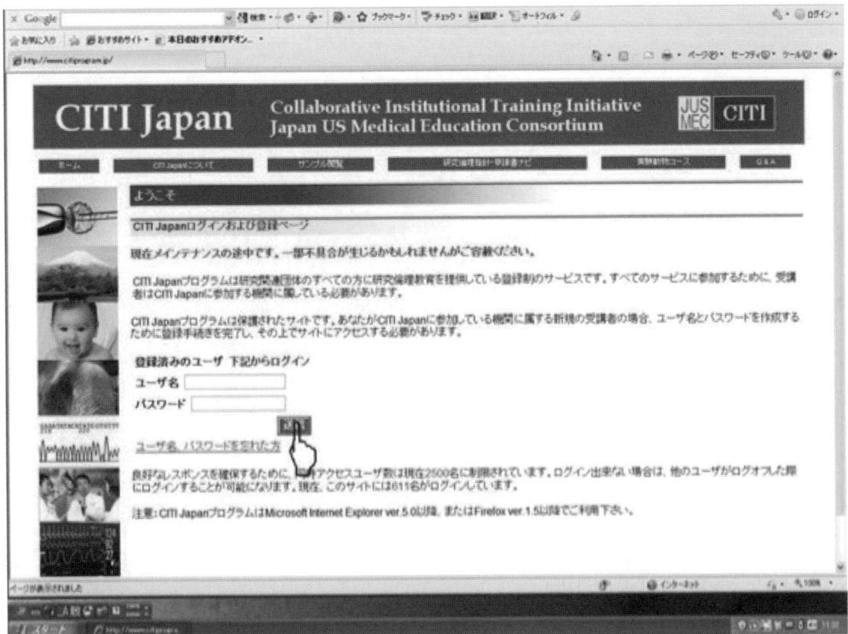

Figure 3. Home page of CITI Japan program. This site (www.CITIprogram.jp) is also accessible with a click from the CITI program (www.CITIprogram.org).

2. Government funding, through academic institutions, has helped to create and run the program.
3. The program is run by a private nonprofit organisation, not any particular academic institution, so that territorial concerns are kept minimal in gaining both expert contributors and user institutions.
4. Particular attention was paid to user-friendliness, through multiple editing processes and the use of graphic illustrations (cartoons) so that users could go over the modules in a relaxed atmosphere at home (Figure 5).

Apart from a number of financial, technical, and personnel issues that we have faced in creating and running the modules, we think that the major hurdles in acquiring users stems from both a lack of an oversight body for education and, again, a scarceness of native experts. The lack of the former fails to give research institutions the incentive to mandate their researchers to use quality education programs. As David Babington-Smith from Epigeum stated, unless mandated, it is unlikely that the researchers will spend time with ethics education materials. Scarceness of

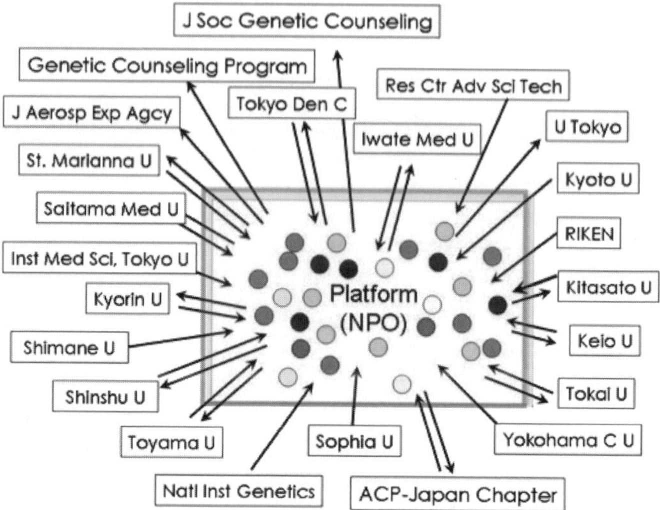

Figure 4. The institutions and professional societies of contributors and users. Although there are not many, we have been lucky to have the few contributors (those with arrows toward the platform) willing to sacrifice their schedules to participate. Institutions who have started subscribing the CITI Japan program (those with arrows from the platform) have research faculty member(s) in leadership roles in research who are aware of the need for ethics education. Various teaching modules are illustrated by coloured circles.

Figure 5. An example of cartoon used in the Japanese modules. Attention was paid to user-friendliness so that the learners feel comfortable going through the modules at home.

experts, particularly those from the science field, results in the inability for institutions to set up necessary programs, the right vehicle for good ethics education. Our program has thus far been funded almost entirely by national government funds. However, as with the CITI program, an ethics education program such as ours should be supported by the users themselves, in a way professional journals are supported by subscribers. It is only recently that the Japanese government started to attach indirect costs to their research funding. Thus, researchers and their institutions have yet to become aware that ethics education of researchers is a perfectly fitting item to be included in those indirect costs.

In summary, although much time may be required for the entire Japanese community to acquire "autonomy" among researchers in both teaching and learning research ethics, we are hopeful that we will continue to be successful as we have been under the unselfish support from growing number of conscientious individuals.

References

1. http://en.wikipedia.org/wiki/Humanitarian_response_to_the_2004_Indian_Ocean_earthquake.
2. http://www.australia-sydney.info/news/20050119tsunami.html.
3. http://www2.ttcn.ne.jp/honkawa/0920.html.
4. http://www.scj.go.jp/ja/info/kohyo/pdf/kohyo-19-t1031-8.pdf.
5. http://www.mhlw.go.jp/general/seido/kousei/i-kenkyu/rinsyo/dl/shishin.pdf.
6. http://www.harvardmacy.org/Programs/Overview.aspx#51.

CHAPTER 39

PROMOTING BEST PRACTICES FOR SCIENTISTS AND POSTDOCTORAL FELLOWS

Makoto Asashima

Currently, intense global information transmission, communication and competition are prevalent in the world of science. Under these circumstances, postdoctoral fellows continuously obtain new information through electronic journals (EJ) and, on the other hand, their studies are assessed through the evaluation of journals (by impact factor, for example). This means that evaluations are made based on something other than the preservation of intellectual property and the traditional quality of academic studies, such that academic positions for young researchers are decreasing, that there is high competition for obtaining research grants, and that the quality and quantity of research papers depend on journals. We publish books so that the young generation of researchers avoids misconduct, as well as to show the enjoyment and importance of academic studies (Figures 1 and 2). We also carry out enlightenment activities by publishing books that show the enjoyment of academic studies through appropriate content. We would like to investigate the problems currently confronted by the world of science and universities, such as the situation of scientists and researchers, the attitudes of researchers ought to be, problems in grant distribution, how post-intellectual property should be, and the relationship between science and society (Figures 3, 4 and 5).

Published Sep. 2007

Figure 1. The young people who aspire to read Sciences Guide book to avoid the misconduct by the Science Ethical Committee.

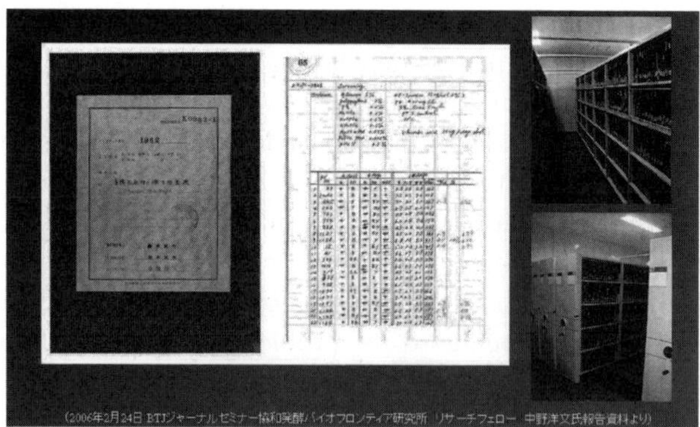

Figure 2. "Blue Note Book" of Kyowa Hakko Co., Ltd., "For the researchers, everything should be written on this notebook including ideas, results and discussions", since July, 1953.

1. Responsibilities of Scientists
2. Conduct of Scientists
3. Continuous Professional Development
4. Accountability and Disclosure
5. Research Activities
6. Establishing Sound Research Environments
7. Compliance with Laws and Regulations
8. Consideration for Research Subjects.
9. Relations with Others
10. Rejection of Discrimination
11. Avoiding Conflicts of Interest
12. Journal Publication Rule
13. Different Situation for Global and Local Region

Figure 3. Code of Conduct consisting of ethical principles.

1. The Responsibility of Organizational Managers
2. The Need for Education on Research Ethics
3. Important Points for Research Groups
4. Important Points for Research Processes
5. Dealing with Misconduct in Research
6. Establishing a Self-monitoring System
7. Freedom with honest mind
8. Reporting fairly to Society and Media
9. Communication and Discussion with Co-Workers

Figure 4. Toward autonomous implementation of the Code of Conduct for scientists.

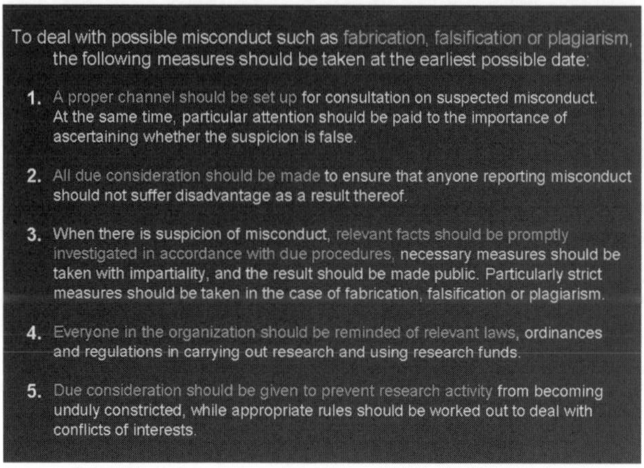

To deal with possible misconduct such as fabrication, falsification or plagiarism, the following measures should be taken at the earliest possible date:

1. A proper channel should be set up for consultation on suspected misconduct. At the same time, particular attention should be paid to the importance of ascertaining whether the suspicion is false.

2. All due consideration should be made to ensure that anyone reporting misconduct should not suffer disadvantage as a result thereof.

3. When there is suspicion of misconduct, relevant facts should be promptly investigated in accordance with due procedures, necessary measures should be taken with impartiality, and the result should be made public. Particularly strict measures should be taken in the case of fabrication, falsification or plagiarism.

4. Everyone in the organization should be reminded of relevant laws, ordinances and regulations in carrying out research and using research funds.

5. Due consideration should be given to prevent research activity from becoming unduly constricted, while appropriate rules should be worked out to deal with conflicts of interests.

Figure 5. Dealing with misconduct in research.

CHAPTER 40

STATENS SERUM INSTITUTE'S COURSE ON GOOD SCIENTIFIC PRACTICE: WHY? HOW? WHAT? DOES IT WORK? WHAT IS NEEDED?

Nils Axelsen

1 WHY?

Denmark has national guidelines for good scientific practice (GSP), and was first in Europe to establish a national body for handling allegations of scientific dishonesty. However, the guidelines are not sufficiently known or used, the importance of systematic and mandatory GSP/RCR (responsible conduct of research) education is not well understood or ignored: Denmark is falling behind. I therefore decided to try and raise the standard at my institution.

2 HOW?

The platform is a newly created full time job as institutional ombudsman for research integrity with the important additional responsibility for educating all my institution's researchers in GSP, strongly backed by the institution's CEO. The first target group are our PhD students and young postdoctorates, hoping that these grassroots will convince their mentors to join in. Format: seminars for 8–10 participants, 5–6 sessions of 3 hours over 14 days. Most topics are presented by the participants and discussed with seasoned prominent medical researchers, the CEO (1 hour in each course), and me. The course is mandatory here within 1.5 yrs after arriving at the institute, credited as a PhD course by the University of Copenhagen.

3 WHAT?

- The modern history of scientific dishonesty: famous cases that caused turmoil and heart-searching
- The roaring world of biomedical science: What constitutes a good paper? The increasing role of science in modern society. Communicating with the public. Growth of numbers of researchers/journals/publications. Stakeholders. Denmark's and SSI's ambitions, options and place in this picture.
- Full commission reports on two Scandinavian cases of suspected dishonesty
- A glimpse of the world of psychopaths and their victims
- The large zone of Questionable Conduct of Research (QCR) and its possible reasons (Anderson *et al.*)
- The Danish Guidelines on Good Scientific Practice
- Dramatic Play on authorship (Macrina), extended with: thorough discussion of the Vancouver rules; how to solve author disputes; responsibility of editors and referees; dilemmas.
- Sorting the good from the bad and ugly (Vaux)
- The ten most important things to know about research ethics (Pimple)
- How to implement GSP/RCR in daily life; counteract QCR/dishonesty; use the ombudsman; be updated.

4 DOES IT WORK?

Yes. It has been a major eye opener: participants feel much better equipped, and suggest that this should be mandatory also at the universities — **and** for their mentors, who now are eager to participate(!). The teachers find the course necessary and timely and enjoy sharing their wisdom, and the participants appreciate meeting them in this small setting. The ombudsman becomes visible, and the education makes his job rewarding. But, of course, there is some way before GSP is satisfyingly consolidated in the institute.

5 WHAT IS NEEDED?

In my opinion, there is a need in Europe for a permanent working party of experienced GSP/RCR educators for sharing ideas, approaches and experiences. Properly organised, my ombudsman model could work at the level of university faculties by making a group of prominent internal senior faculty responsible for the discussion part of the courses.

CHAPTER 41

RESPONSIBLE CONDUCT OF RESEARCH WORKSHOPS AT THE AUSTRALIAN NATIONAL UNIVERSITY

Simon Bain

It is recognised that for institutions to develop a strong institutional culture with regard to research integrity, there is a need for a combination of a necessary "top down" approach, which entails strong support from the university executive, together with a concomitant "bottom up" approach.

Relevant to the latter, the Australian National University made the decision to run mandatory Higher Degree Research Responsible Conduct of Research workshops for PhD and Masters students, commencing in 2009, and this had the strong backing of members of the ANU Executive. The workshops are run by the ANU's Office of Research Integrity across the University's seven colleges with some change in presentation format between colleges in recognition of research discipline differences in accepted practice. Students must sign off as having attended a workshop by the completion of the first year of their higher degree. Much of the workshop content is based on the Australian Code for the Responsible Conduct of Research and related ANU policy, but significant reference is made to recent international research relative to the responsible conduct of research, the concept of questionable research practices (QRPs), and reference to high profile cases of research misconduct.

The format consists of two intensive one hour sessions separated by a 15-minute break. The first session initially covers recent research findings relevant to the components that make up research integrity, followed by an outline of the principles central to the responsible conduct of research: authorship, management of research data, supervision of research trainees, publication and dissemination of research findings, peer review, conflicts of interest, and collaborative research. The second session begins with a discussion on research ethics, including the need, the process, and situations that might arise. The session continues with a discussion on questionable research practices and concludes with examples of and a discussion about research misconduct.

A small number of case studies are presented through the presentation and student interaction is encouraged at all times. The presenter is frequently impressed by the depth of perception that some higher degree research (HDR) students exhibit relative to the responsible conduct of research.

The presenter's experience relative to the workshops has been one principally of satisfaction gained from student engagement. It is felt that these workshops do develop a bottom up appreciation of the significance of having a responsible research culture in place.

For the past three years, similar workshops have been run as part of a Future Research Leaders Program at the University. The format is broadly the same as the HDR workshop with some change in emphasis. This program engages principally with researchers at a mid-career level, the number of participants is relatively small compared with the HDR program, but engagement with researchers at this level enables some ability to assess how we are progressing with promoting a responsible research culture within the University.

CHAPTER 42

SCIENTIFIC INTEGRITY: THE PERSPECTIVE FROM IMPERIAL COLLEGE LONDON

Mary Ritter and Stephen Webster

When Imperial College set up its Graduate Schools ten years ago, tasked with supporting PhD students with the broader aspects of their studies, thought was immediately given to the question of how we should teach ethics.

Now, in 2010, the Graduate Schools are as established a part of the College landscape as the student bar. But when it was being set up, and students were being asked to attend half-day courses that took them away from their daily laboratory routines, everyone involved knew that some persuasion would be necessary. It was necessary to make obvious the immediate and significant utility of any course we chose to offer. That is not too difficult for communication classes, or statistics support, or "How to Write Your Thesis". But ethics?

It is not that science institutions are oblivious of ethical discussion. On the contrary, Imperial College takes very seriously its ethics committees which adjudicate on clinical trials. And for some time, medical ethics has been an established part of the Medical School's undergraduate curriculum. Overall, however, for a research scientist, the field of ethics is important simply as an onerous aspect of the management of some — certainly not all — scientific projects.

It is often said that scientific integrity is hard to define and, because of its breadth, difficult to teach.[1] Yet, as the Graduate Schools developed, the

[1]Responding to challenges in educating for the responsible conduct of research. Kallichman, M. W. (2007). *Academic Medicine*, 82(9): 870–875.

scientific community's interest in research ethics was growing. Exploding on the scene were high profile cases of misconduct, in particular the cases of Jan Hendrik Schön in 2002[2] and of Woo Suk Hwang in 2005. These cases no doubt relate to questions of character, but they also encourage reflection on the process of peer review, on the regulation of science, and on the way science is represented in the media.[3]

Also in these early years of Imperial's Graduate Schools, there was in the pages of *Nature* magazine an interesting debate concerning the question of young scientists getting fair credit for their work.[4] It seemed that as every week passed, an editorial or news item in *Nature* magazine confirmed scientific integrity as a new and powerful theme for scientists everywhere to reflect upon. We were aware that in the USA the debate was forging ahead, with real problems emerging about how broadly the definition of misconduct should be set. David Korn, Senior Vice-President of the Association of American Medical Colleges, had suggested that while the federal regulatory mechanisms should "... be circumscribed and focused on transgressions that are reasonably unambiguous and are unacceptable across all scientific and scientific disciplines", within the scientific community itself, there would be continuing debate about the boundaries between ethical and unethical behaviour.[5] This was significant, not only because of Korn's desire to keep the federal remit as simple as possible, but also because of the clear indication that scientists themselves need to work out for themselves where the limits lie, and where ambiguities remain. As we ourselves set to thinking how to teach research ethics, it did not look likely that instruction would be a matter of simply teaching a set of rules. On the contrary, what was interesting about all this was that a topic of vital importance seemed to be generating many views in the community, some of them contradictory. Were not these signs of disagreement a very promising resource for our teaching strategy? Surely our young scientists, so vital a part of our

---

[2]Reflections on scientific fraud. Editorial. *Nature* (2002), 419, 417.

[3]Modest witnessing and managing the boundaries between science and the media: A case study of breakthrough and scandal. Haran, J. and Kitzinger, J. (2009). *Public Understand. Sci.*, 18(6): 634–652.

[4]Rank injustice. The misallocation of credit is endemic in science. Lawrence, P. A. (2002). *Nature*, 415: 835–836.

[5]Scientific misconduct: The state's role has limits. Korn D. *Nature*, 420, 2002, 739.

community, would want to take part in those reflections, and would have their own voice?

Fortunately, Imperial's Graduate Schools had invested in classes in research integrity with institutional significance by rewarding attendees with a contribution to the credit score necessary for transfer from the MPhil to the full PhD degree. With such institutional backing, classes could be regular and could expect to recruit 20 or more students. Now we could experiment with the pedagogy.

Keynote lectures we had tried. We discarded them. Lectures about research ethics are efficient in getting information across, in sounding warnings, and in telling picaresque tales of promising scientists getting into trouble. However, through their format lectures amplify two distortions especially problematic for ethics teaching. Firstly, it is difficult for the lecturer to avoid suggesting he or she knows the difference between right and wrong, whereas the audience at the lecture do not. Secondly, a related point, the audience are passive recipients. Not only was discussion between students difficult to manage in such an environment, it seemed impossible to draw from the students any contributions based on their own experience.

We were helped by the burgeoning literature on scientific integrity, and texts suitable for students were appearing.[6] In the UK, annual reports on publication ethics were being prepared by a group of journal editors, each edition containing case material and reflective articles, and appearing annually.[7] The various codes of conduct associated with various branches of the sciences all revealed the complexities of the issue, and made good material for discussion. In March 2007, the College welcomed Sir David King, then chief scientific advisor to the British government, to a meeting of graduate students and staff where he launched a new code he had himself initiated, named *Rigour, Respect, Responsibility: A Universal Ethical Code for Scientists.*[8]

There was, then, no shortage of teaching material. We could talk about the main types of misconduct (fabrication, falsification, plagiarism) and we could mention too the more nuanced areas of ethical heartache, such as issues to do with authorship, peer review, collaboration and data management. Attempts had been made to measure the amount of misconduct

[6] *The Ethics of Science* (1998). Resnick, D. B. Routledge.
[7] http://publicationethics.org/. Accessed on 10/10/10.
[8] http://www.berr.gov.uk/files/file41318.pdf. Accessed on 10/10/10.

occurring[9] and we could describe codes of conduct and their likely impact.[10] We could remind the students of the grave consequences of infringements of scientific norms,[11] and we could re-tell those colourful stories of scientific disgrace we had used to spice our lectures.

Useful though these resources are, none of them are necessarily engaging or relevant for a student. We saw that, in terms of student involvement, a workshop can be as mute as a lecture. We accepted of course the sound advice that workshops can be made more dynamic by providing the students with case studies, those collages of scientific sharp practice. Case studies do get the students talking, we found. Yet these stories are sometimes far-fetched. Anonymised and constructed, they do not necessarily "touch" the student. We needed material that was more immediate. Could the students themselves fill the gap?

We noticed that, paradoxically, the more dramatic and disastrous our examples, the quieter the class. Conversely, the milder the problem, the more engaged were our students and the livelier the discussion. The trick seemed to be to construe scientific integrity as an everyday affair, a matter of how we deal with each other in a competitive, and pleasantly unregulated environment. When we implied that scientific integrity was an issue involving deluded scientists running amok, then our students' eyes clouded over.

Mostly we deal with first year PhD students. They have ideals, and they have not yet been roughened by a tough environment. Yet they know that money is important, and so is publishing, that laboratories vary enormously, that luck helps, and that a supportive supervisor makes all the difference. And in a group of 15 such students, two or three will have a stronger story to tell. With such material, a teacher can choreograph three hours of valuable work.

Every session, we get the ball rolling by giving the assembled students a simple task, "Tell the person sitting next to you a good thing, and a bad thing, that has happened to you in science." What emerges is a mix of gripes

---

[9]Marshall, E. (2000). How prevalent is fraud? That's a million-dollar question. *Science*, 290: 1662–1663.

[10]Scientific societies and promotion of the responsible conduct of research: Codes, policies and education. Macrina, F. (2007). *Academic Medicine*, 82(9): 865–869.

[11]Couzin, J. (2006). Scientific misconduct: Truth and consequences. *Science*, 313: 1222–1226.

about experiments not working and time seeming short, and hosannas about the pleasures of the craft, of getting results, of perhaps getting towards a publication. Simple stuff, you might say, but it is central to the students' experience. And it is permeated with a realisation, as the discussion proceeds, that these small and intimate aspects of a student's research are actually very fragile, and must be held firm by a culture that takes its integrity seriously. Students learn, in other words, that scientific integrity, far from revolving around dramatic and far-away felonies, is a matter for themselves, and their career, now. In that understanding, the most important resources, and the most important instructors, are the students themselves.

CHAPTER 43

WORKSHOP #3 REPORT: INTERNATIONAL RESPONSIBLE CONDUCT OF RESEARCH EDUCATION[1]

Nicholas H Steneck, Mike Kalichman and Nils Axelsen

20 February 2011

1 SUMMARY

With support from the US National Science Foundation, a workshop on International Responsible Conduct of Research Education was organised in conjunction with the 2nd World Conference on Research Integrity, Singapore, 21–24 July 2010.

The immediate goal of the workshop was to identify and bring together an international group of responsible conduct of research (RCR) educators and other interested parties to address the need for improved consistency and availability of RCR education for students and researchers on a global level. This general goal was divided into four specific aims:

- Aim 1. Identify and bring together an international group of RCR professionals.

[1]The workshop was partially supported by a grant from the US National Science Foundation (0940073, International Responsible Conduct of Research Education Workshop). For a copy of the full final report, dated 20 February 2011, contact nsteneck@umich.edu. The full report is available at www.singaporestatement.org/nextsteps.html/

- Aim 2. Develop international guidelines for RCR education.
- Aim 3. Develop plans for the circulation and adoption of the international RCR guidelines.
- Aim 4. Develop plans for expanding the international community of RCR professionals.

Overall, this effort sought to improve integrity in research and the benefits this brings, such as better protection for human subjects, less waste and careless practice, enhanced objectivity, and a sincere interest on the part of researchers in working for the greater public good.

Based on the Workshop discussions and subsequent discussion, four general conclusions are proposed as an aid to future work:

1. Addressing the responsible conduct of research from an international perspective is essential. Research is increasingly collaborative across international borders, and research professionals frequently work in countries other than where they were trained.
2. Defining the parameters for responsible conduct of research education is best accomplished through multinational conversations. A focus on the details of specific questions and approaches is essential for clarifying areas of agreement.
3. Translating the results of these conversations so that they are linguistically, culturally, and socially appropriate in particular international settings is the only way to ensure meaningful results across borders.
4. Effective education depends on widely disseminating the content and approaches that have been found to be most useful. This could best be accomplished through an ongoing series of international, train-the-trainer workshops.

Efforts are currently underway to disseminate and build on the workshop recommendations. Funding from this project supported travel for PI Steneck to present a series of talks on RCR and research integrity at the First Brazilian Meetings on Research Integrity (December 10–16, 2010). The recommendations were also presented to the European Commission, the European Research Council, and COST (European Cooperation in Science and Technology) in January 2011 as part of reports on the 2nd World Conference on Research Integrity. Additional efforts that grow out of this project and other related efforts will form the basis of continued discussion of RCR education during the 3rd World Conference on Research Integrity, which is currently in the planning process.

2 STATEMENT OF NEED

The need for formal RCR education became widely recognised in the USA in the 1990s, following publication in 1989 of the Institute of Medicine Report, The Responsible Conduct of Research in the Health Sciences and the National Institutes of Health Training Grant requirement. Since that time, a wide range of courses, textbooks, other teaching resources, and programs have proliferated in the USA, with the end result that most research institutions now offer some formal RCR training. The quality of this training varies considerably, even within institutions, but the fact remains that the need for some formal RCR education is a recognised part of research training in the USA, as evidenced most recently by the Congressional mandate in the 2007 America Competes Act and response by the National Science Foundation.

The need for formal RCR education is not widely recognised internationally. Countries that have in one way or another addressed the issue of research integrity, frequently mention the importance of RCR education, but to date no countries have followed the lead of the USA and adopted formal requirements. Consequently, formal RCR education is not available to most research students or new researchers outside the USA.

Uneven and inconsistent RCR education poses a significant challenge for international research collaboration. As more and more governments encourage researchers to work with colleagues in other countries, either virtually or in person, clear and consistent guidelines for responsible practice are essential. Researchers must know what is expected of them professionally when they work in and with colleagues in different countries. Moreover, to avoid confusion and to promote efficiency in research, the guidelines for responsible practice should be harmonised across national boundaries. Scientific laws and theories are not country dependent. The same should be true for the guidelines and best practices for responsible research. The global Singapore Statement on Research Integrity (see Appendix of this book and also www.singporestatement.org) adopted at the 2nd World Conference on Research Integrity was formulated as a way of promoting discussions on the harmonisation of research principles and responsibilities.

3 OUTCOMES

3.1 *Aim 1. Identify and Bring Together an International Group of RCR Professionals.*

As planned, the workshop was held the day following the conclusion of the 2nd World Conference on Research Integrity (24 July), in the Pan Pacific Hotel in Singapore. The 49 attendees represented 19 countries. The workshop program, resources, handouts, and attendance summaries are included in Workshop Report (see p. 275, footnote 1).

3.2 *Aim 2. Develop International Guidelines for RCR Education.*

The career and educational path for researchers varies from country to country. To bring more uniformity to researchers' understanding of their responsibilities as researchers, workshop participants addressed and reached general agreement on five key elements of the planning and delivery of RCR education:

3.2.1 *Goals*

The primary overarching goal of RCR education should be to promote the acquisition of the knowledge and skills that will enable researchers and research institutions to understand and follow the fundamental principles of responsible research, such as the ones outlined in the Singapore Statement. In more specific terms, RCR education should:

- Promote knowledge, skills and attitudes that enhance research integrity across all disciplines and at all levels of the research enterprise.
- Develop programs and opportunities that help students, researchers and others act in responsible ways as researchers.
- Develop programs and opportunities that enable students, researchers and others to reflect upon and discuss research ethics, ethics in science, responsible conduct of research and good research practices.
- Enhance the long-term professional development of present and future researchers.

3.2.2 *Audience*

RCR education should be provided for everyone engaged in research. It can and ideally should be organised by broad research fields, research roles and career stage. Adopting a single approach for everyone engaged in research is not the most effective way to provide RCR education.

3.2.3 *Knowledge and Skills*

RCR education programs and courses should provide both knowledge (what researchers should know) and skills (what they should know how to do). The content and organisation of courses and programs will vary with different audiences. The following summary describes major topics that at some level, and as appropriate to particular audiences, should be included in courses and programs:

A. Research as a Profession

1) Importance of integrity and responsibility in research
2) Conflict of interest and commitment
3) Collaboration
4) Mentoring
5) Social responsibility
6) Whistleblowing

B. Research Record

1) Research design and data management
2) Data sharing
3) Intellectual property

C. Credit and Responsibility

1) Authorship
2) Publication
3) Peer Review

D. Regulations

1) Misconduct
2) Human and animal subjects
3) Intellectual property
4) Laboratory safety
5) Global regulations

E. Research and Society

1) Global responsibilities
2) Different cultural and national perspectives
3) Dual-use research

F. Skills

1) Asking questions
2) Where to find more information
3) Ethical decision making
4) Conflict resolution
5) Leadership and management skills

3.2.4 *Approaches to Implementation*

RCR education must be carefully planned and implemented if it is to be successful. Anyone planning such an RCR program should consider the following points:

A. General Considerations

1) The elements (courses, mentored experiences, and other) must be relevant and designed with the full education program in mind.
2) Courses, programs, and approaches should be designed in a way that allows meaningful assessment.
3) Some training should take place in research settings through mentoring, modelling good practices.
4) RCR education can be distributed across training and embedded in other elements of research training.
5) Major emphasis should be placed on good research practices.
6) For an RCR education program to be successful, it is necessary to educate the RCR educators (train the trainers).

B. Formats

1) Traditional learning settings: courses, seminars, lectures and so on
2) Web-based teaching

C. Discussion tools

1) Engaged (active) learning is essential.
2) Case studies are usually a first choice.
3) Many other discussion tools might be used, such as role playing, debates, student teaching, and videos.
4) Using different methods is optimal.

D. Institutional RCR Program Development

1) Strong leadership is necessary.
2) Program development can be either top-down or bottom-up, depending on the institution.
3) Having or creating an enabling environment is important.
4) Program development must be monitored and assessed.

3.3 *Aim 3. Develop Plans for the Circulation and Adoption of the International RCR Guidelines.*

This Workshop Report will be published in *Promoting Research Integrity in a Global Environment*, which contain papers and reports from the 2nd World Conference on Research Integrity, and on the Singapore Statement web site (www.singaporestatment.org) (www.singaporestatement.org/nextsteps.html/). Copies will also be sent to workshop participants for further distribution and made available upon request.

3.4 *Aim 4. Develop Plans for Expanding the International Community of RCR Professionals.*

A number of options were discussed at the end of the workshop for continued discussion and next steps, including:

1) Organise an international RCR train-the-trainer workshop.
2) Engage the co-operation and help of professional/scientific/academic societies and professional organisations, such as university associations.
3) Target key agencies within countries/agencies, working with and through professional societies and professional organisations to encourage the development of training guidelines/standards/recommendations.
4) Develop mechanisms for assessing progress on a global basis, with the goal of international accountability.
5) Co-opt respected researcher leaders to champion RCR education.
6) Promote the development of nationally appropriate training and materials following the framework recommended above.

To make continued progress in promoting RCR at the international level, future discussions need to focus on three crucial topics:

1) Research Misconduct:
   What are the research behaviours that are clearly unacceptable whether or not already prohibited by nation-specific regulation?
2) Research Standards:
   What are the areas of greatest likelihood for misunderstandings, disputes, and/or possible misconduct?
3) Tools to Encourage Discussion:
   How can the list of approaches for encouraging discussion in various venues be expanded and evaluated, including the research environment, in the curriculum, in courses dedicated to RCR, and online?

Finally, it is strongly urged that as discussion continues, efforts should be made to convene one or more international train-the-trainer workshops. These workshops should be intensive, 3-day programs, jointly taught by two or more faculty of different nationalities. The spirit of the workshops should be based on multinational participation, and a sharing of different approaches and understandings — not just a didactic presentation of a fixed curriculum.

To further these efforts and to build on the 2nd World Conference on Research Integrity RCR Education Workshop, the organisers plan to: (1) develop a contact list to keep the global RCR education community in touch with one another, (2) post related events and publications on the Singapore Statement web site, and (3) assure that RCR education is included in planning for the 3rd World Conference on Research Integrity.

SECTION VII

INTEGRITY ISSUES FOR AUTHORS AND EDITORS

INTRODUCTION

As gatekeepers for the research record, editors can play an essential role in promoting integrity and responding to misconduct in research. To responsibly fill their role as gatekeepers, Kleinert argues that editors need to be well trained, operate from clear principles, and seek to motivate and educate others on responsible practices. From the perspective of a leader of the Committee on Publication Ethics (COPE), she suggests that editors need to adopt and work from common principles and responsibilities, as well as collaborate with colleagues. This is particularly important when responding to allegations of research misconduct, since editors seldom have the resources to investigate cases themselves. COPE's efforts to provide better and more uniform guidance for editors is described in more detail by Wager, such as the COPE Code of Conduct, Best Practices, and flowcharts that provide guidance for handling difficult research misconduct cases.

EQUATOR has taken on the special challenge of enhancing the quality of health research through the dissemination of information and training. Numerous guidelines already exist for best practice in clinical research, such as the CONSORT Statement, PRISMA, STROBE, STARD, or SQUIRE. Simira and Moher suggest, however, that guidelines may not be enough. Researchers often are not aware of them, do not know how to use them, or may not have the incentives needed to do so, leading to the suggestion that clearer policies and requirements are needed.

Some of the abuses of best practice that can emerge are illustrated by the final two articles in this section. Arnold describes a flagrant case of plagiarism that took many months to resolve, without any co-operation being received by the publisher of the article that contained the plagiarism. He wonders whether profit considerations stood in the way of prompt and appropriate action by the publisher. That the publishers and editors of journals are not above reproach is illustrated by Arnold's second example, the use editors make of questionable practices to inflate their journal's impact factor. Vlassov describes another problem that exists for accuracy and reliability in publication, namely, the problem of lingering political influence and inappropriate deference to well-established, senior colleagues. His closing challenge is to the academies of science and funding bodies in Russia and other countries to set up committees to look into these problems.

Two COPE-initiated standards documents at the end of this Section, provide global guidance on *Responsible Research Publication* in the form of *International Standards for Authors* and *International Standard for Editors*.

CHAPTER 44

CHALLENGES FOR EDITORS AS GUARDIANS OF THE RESEARCH RECORD

Sabine Kleinert

Editors are in a unique position as guardians and stewards of the research record to both indirectly foster and encourage research integrity and to react appropriately to suspected research and publication misconduct. With publication being the generally accepted currency in the research world for recognition, reward, and promotion, this is a powerful position, which gives editors an important role as leaders in the field of research integrity in the wider sense and of publication ethics in particular.

Leading means first and foremost to have a goal and a vision. For editors, the main and overarching goal should be to encourage and ensure full and honest reporting of research that needed to be done and was conducted responsibly. The next step is to set a path to achieve this goal in the form of editorial principles, best practices, and policies. Finally, as leaders editors should educate and motivate others, including peer reviewers, authors and editors in all fields, editors of journals large or small, general or specialised, to follow best practice.

What are the main challenges when there is alleged or suspected misconduct? First, and this should not be forgotten or underestimated, editors themselves need to be educated. Raising awareness and guiding editors is the major purpose of the Committee on Publication Ethics (COPE). By issuing guidelines, flowcharts, and advising on real cases, COPE aims to educate and help editors in dealing with issues of research and publication ethics. A second challenge is to agree on common principles and approaches among authors and editors. The maximum effect and influence on responsible research conduct and responsible publication can only be achieved if all journals and editors work with the same principles and goals. We have made a first attempt to outline and agree on, among participants at the conference in Singapore, international standards for authors and for editors (see Section 7, below). The third challenge, and perhaps most difficult challenge, is a common agreed and improved collaboration between editors, funders, and institutions.

With the exception of plagiarism and duplicate or redundant publication, editors cannot themselves investigate allegations of research or publication misconduct but must rely on institutions or national bodies to conduct investigations. From my experience at *The Lancet* and at COPE, there are many difficulties for editors when collaborating with institutions. It is often not clear whom to contact. When contacted, those responsible may not respond. Some forms of misconduct, such as self-plagiarism or reviewer misconduct, may not be taken seriously. Investigations may not be done at all, may not be done thoroughly or fairly, and may take a long time. Editors are not always informed about investigations or the outcome of investigations. Findings may not be made available publicly.

As such, one of COPE's next planned activities is to draw up a guidance document on the ideal collaboration between editors and institutions. Only if all involved in the research enterprise will work together and act as both examples and leaders, will research be trusted and retain its true purpose: to advance human knowledge.

CHAPTER 45

PROMOTING INTEGRITY IN RESEARCH REPORTING: DEVELOPING UNIVERSAL STANDARDS AND PROMOTING BEST PRACTICE AMONG JOURNALS

Elizabeth Wager

Integrity in reporting research is important because publications form the basis for future research decisions and the application of findings. Therefore, inaccurate or misleading publications may lead to inappropriate policy decisions and, ultimately, public harm.

In most research fields, peer-reviewed journals are an important vehicle for research reporting, and editors of such journals therefore have important responsibilities for maintaining the quality and integrity of the research record. Journals can play a role in promoting research integrity and responsible reporting by educating authors and reviewers. Editors should adopt policies and practices that act as deterrents against, or help to detect, misconduct. Editors have a responsibility for correcting the literature, e.g. by retracting seriously flawed articles.

However, while journals, and their editors, can contribute to integrity in reporting, they may also contribute to misconduct. For example, reviewers may be biased, may not disclose competing interests or may seek to delay publications by rivals or even plagiarise. Similarly, editors may abuse their position, exhibit bias, publish their own work without proper review, or fail to disclose competing interests. Lastly, publishers and journal owners, such as academic societies, may interfere with editorial decisions for commercial or political reasons.

Although journal editors carry heavy responsibilities for ensuring proper processes in peer review and safeguarding the integrity of the literature, they rarely receive any training in such issues. Most academic editors fulfill their editorial duties part-time, in addition to their regular jobs, and may receive minimal support from the journal's publisher or owner. Organising peer review and journal production is time-consuming; editors may therefore not have time to review their journal's policies or ensure they are following best practices.

Several organisations exist to help editors and to promote good practice in publication. These include the Committee on Publication Ethics (COPE), the Council of Science Editors (CSE), and the World Association of Medical Editors (WAME). COPE is unique in focusing solely on publication ethics and also encompassing all academic fields (not just science or medicine). It has published a Code of Conduct for editors, which it expects its members to follow, and Best Practice guidelines, which set more aspirational targets. To help editors follow the code, it has also produced a series of flowcharts covering the most common ethical issues such as redundant publication, suspected plagiarism, reviewer misconduct, unethical research and authorship problems. It has also produced specific guidance on topics such as retraction. These are freely available on the COPE website (www.publicationethics.org).

For questions that cannot be answered by the flowcharts or guidance documents, COPE provides a quarterly forum for members to discuss anonymised cases. COPE does not investigate cases nor provide formal adjudication, but provides informal advice based on cases presented by journal editors or publishers. After discussion, the cases, the advice from the meeting, and a report on the outcome (if provided by the editor) are entered into a searchable database that now contains over 300 cases.

COPE was founded by the editors of several prominent British biomedical journals but now has over 6,000 members from all over the world and from a wide range of academic fields. We have therefore been reviewing and revising our documents and guidance to ensure they are applicable across all disciplines and regions. While the main forum meetings still take place in London, members can present cases via phone, and we have also held a forum in Singapore (associated with the World Conference on Research Integrity) and plan another to be held in the USA in late 2010.

COPE's guidance appears to be widely appreciated by editors and publishers. The flowcharts have been translated into several languages including Chinese, Japanese, Korean, Spanish, and Turkish. We are currently developing a distance learning package, a short guide to publication ethics for new editors, and a code of conduct for publishers. We hope that these, and other initiatives, will ensure that editors and publishers of academic journals can meet their important responsibilities and become increasingly effective in preventing, detecting and responding to research and publication misconduct.

CHAPTER 46

THE EQUATOR NETWORK: A GLOBAL INITIATIVE TO IMPROVE REPORTING OF HEALTH RESEARCH STUDIES

Iveta Simera[1] and David Moher[2]

Providing a complete, accurate and clear account of conducted research studies in scientific publications is an integral part of responsible research. Yet the literature is full of examples documenting inadequacy of health research reporting: not publishing whole studies or selecting only some outcomes for publication with "attractive" results; inadequately described methods and interventions preventing their assessment and replication; confusing or misleading presentation of results, data, graphs, and images; or inadequate reporting of harms, which in particular can have serious consequences for patients' safety [1]. These and other reporting problems undermine the reliability of published research and seriously limit the usability of findings in clinical practice and further research. Improper practices also compromise the returns on large financial investments in health research and waste the time and goodwill of human participants in such studies [2, 3].

In order to raise standards in health research reporting and help scientists, peer reviewers and editors in producing high quality research papers, a number of reporting guidelines have been developed over the past 15 years.

[1]Centre for Statistics in Medicine, University of Oxford, Oxford, UK

[2] Clinical Epidemiology Program, Ottawa Hospital Research Institute, Ottawa, Canada; Department of Epidemiology and Community Medicine, Faculty of Medicine, University of Ottawa, Ottawa, Canada

Author for correspondence: Iveta Simera, iveta.simera@csm.ox.ac.uk

These guidelines specify, usually in the form of a checklist, a minimum set of information needed for a complete and clear account of what was done and what was found in a research study [4, 5]. The most influential so far has been the CONSORT Statement [6] for reporting randomised controlled trials. Other guidelines include PRISMA for reporting systematic reviews and meta-analyses [7], STROBE for observational studies [8], STARD for diagnostic accuracy studies [9], or SQUIRE for quality improvement studies [10]. A complete catalogue of guidance documents can be found elsewhere [4]. Despite the availability of good guidance, many scientists are not aware of their existence and many journals do not instruct their authors to consult and follow these guidelines.

To improve dissemination of reporting guidelines and provide easy access to available guidance supporting good research reporting we set up the EQUATOR Network programme. EQUATOR (Enhancing the QUAlity and Transparency Of health Research) is an international initiative that aims to improve the reliability and value of the scientific literature by promoting responsible reporting of health research [1]. Major EQUATOR activities focus on providing resources through the Network's online portal and supporting the use of these resources through training and educational activities. The EQUATOR website (http://www.equator-network.org/) hosts regularly updated resources for researchers (e.g. guidelines on reporting, scientific writing, ethical conduct in research and publication, etc.), for editors and peer reviewers to support implementation of good research reporting in journals, and for scientists engaged in the development of reporting guidelines to support effective development of high quality guidance.

Our training activities focus on journal editors, peer reviewers, and researchers — authors of scientific papers. Although editors and peer reviewers play an important role in safeguarding the final quality of published research and they need to have a good knowledge of principles of good research reporting, primary responsibility for research and its publication lies with researchers. It is therefore important to target education and training efforts at research students and young starting scientists to help them develop good reporting habits, which can be then maintained throughout their carriers.

An important challenge for the EQUATOR team is to ensure a global awareness of problems associated with poor research reporting and motivate wider use of available tools for achieving high standards of published reports. Many institutions around the world are now developing codes for responsible research conduct. These mostly contain principles and main responsibilities. Using resources collated and provided by EQUATOR helps to translate these principles into practice and apply them in everyday research. EQUATOR tries to ensure that its activities are global and engage a wide range of supporters — journals, research funders, universities and other relevant individuals and organisations. Thus far we have organised workshops and lectures in the UK, Canada, the USA, Thailand, Singapore, and India.

One of our recent major steps in the international promotion of responsible research reporting has been establishing formal collaboration with the Pan American Health Organization/World Health Organization (PAHO/WHO). This partnership will focus on raising standards of research reporting in the Americas and will in particular help to reach non-English native speaking researchers. The first collaborative project involved translating the EQUATOR online resources into Spanish and developing the Spanish EQUATOR website (http://www.espanol.equator-network.org/). Translating important guidelines and explanatory educational papers into languages local to scientists should ease understanding and use of the guidelines. We are also planning to expand these activities into other languages.

Evidence suggests that adherence to reporting guidelines improves the reporting quality of published papers [11–13]. Reporting guidelines, together with other important mechanisms such as mandatory registration of clinical trials, have great potential to reduce poor reporting practice and institute clarity, completeness, and minimum bias in the presented health research.

However, in order to achieve a global improvement across health research literature and maintain high standards, we need clear policies on responsible research reporting, requesting the use of appropriate tools and mechanism and ensuring compliance of all parties involved in research planning, conduct, and publication. The EQUATOR Network aims to actively contribute towards this process and seeks champions and strong partners to join us in promoting these goals.

1 AUTHORS' COMPETING INTERESTS

Dr Moher is supported by a University of Ottawa Research Chair and is a member of the EQUATOR Steering Group. Dr Simera is supported by the EQUATOR programme funding.

2 ACKNOWLEDGEMENTS

The EQUATOR Network is supported by the UK NHS National Institute for Health Research, UK Medical Research Council, Scottish Chief Scientist Office, Canadian Institutes of Health Research, and Pan American Health Organisation. We would like to thank to these funding bodies for providing vital financial support without which this programme would not be able to start and continue its development.

None of the above funders influenced in any way the content of this article.

References

[1] Simera, I., Moher, D., Hirst, A., Hoey, J., Schulz, K. F. and Altman, D. G. (2010). Transparent and accurate reporting increases reliability, utility, and impact of your research: Reporting guidelines and the EQUATOR Network. *BMC Med.* 8(1): 24.

[2] Chalmers, I. and Glasziou, P. (2009). Avoidable waste in the production and reporting of research evidence. *Lancet*, 374(9683): 86–89.

[3] Duff, J. M., Leather, H., Walden, E. O., LaPlant, K. D. and George, T. J. Jr. (2010). Adequacy of published oncology randomized controlled trials to provide therapeutic details needed for clinical application. *J Natl Cancer Inst.* 2010; 102(10): 702–705.

[4] Simera, I., Moher, D., Hoey, J., Schulz, K. F. and Altman, D. G. (2010). A catalogue of reporting guidelines for health research. *Eur J Clin Invest*, 40(1): 35–53.

[5] Moher, D., Schulz, K. F., Simera, I. and Altman, D. G. (2010). Guidance for developers of health research reporting guidelines. *PLoS Med*, 7: e1000217.

[6] Schulz, K. F., Altman, D. G. and Moher, D. for the CONSORT Group (2010). CONSORT 2010 Statement: Updated guidelines for reporting parallel group randomised trials. *BMC Med*, 8: 18.

[7] Moher, D., Liberati, A., Tetzlaff, J. and Altman, D. G. (2009). The PRISMA Group. Preferred reporting items for systematic reviews and meta-analyses: The PRISMA Statement. *PLoS Med*,6(7): e1000097.

[8] Von Elm, E., Altman, D. G., Egger, M., Pocock, S. J., Gotzsche, P. C. and Vandenbroucke, J. P. (2007). The strengthening the reporting of observational studies in epidemiology (STROBE) statement: Guidelines for reporting observational studies. *Ann Intern Med*, 147(8): 573–577.

[9] Bossuyt, P. M., Reitsma, J. B., Bruns, D. E., Gatsonis, C. A., Glasziou, P. P., Irwig, L. M., Lijmer, J. G., Moher, D., Rennie, D. and de Vet, H. C. (2003). Towards complete and accurate reporting of studies of diagnostic accuracy: The STARD initiative. Standards for Reporting of Diagnostic Accuracy. *Clin Chem*, 49(1): 1–6.

[10] Davidoff, F., Batalden, P., Stevens, D., Ogrinc, G. and Mooney, S. (2008). Publication guidelines for quality improvement in health care: Evolution of the SQUIRE project. *Qual Saf Health Care*, 17 Suppl 1: i3–i9.

[11] Plint, A. C., Moher, D., Morrison, A., Schulz, K., Altman, D. G., Hills, C. and Gaboury, I. (2006). Does the CONSORT checklist improve the quality of reports of randomised controlled trials? A systematic review. *Med J of Austr*, 185: 263–267.

[12] Smidt, N., Rutjes, A. W., van der Windt, D. A., Ostelo, R. W., Bossuyt, P. M. *et al.* (2006). The quality of diagnostic accuracy studies since the STARD statement: Has it improved? *Neurology*, 67: 792–797.

[13] Prady, S. L., Richmond, S. J., Morton, V. M. and MacPherson, H. (2008). A systematic evaluation of the impact of STRICTA and CONSORT recommendations on quality of reporting for acupuncture trials. *PLoS ONE*, 3: e1577.doi:10.1371/journal.pone.0001577.

CHAPTER 47

CHALLENGES AND RESPONSES IN MATHEMATICAL RESEARCH PUBLISHING

Douglas N Arnold

1 A CASE OF PLAGIARISM

The world of scholarly publishing is increasingly challenged by journals that do not uphold basic standards of scholarly integrity. Although they typically claim to implement a peer review process, often the process is non-existent or grossly insufficient. One outcome is that plagiarised material is published.

In 2009, the Society for Industrial and Applied Mathematics (SIAM), a professional organisation that publishes about 15 respected journals, encountered such a case when the authors of one its articles found their 175-word abstract, verbatim, on the web page of another journal attached to a paper with a different title and different authors. SIAM set out to investigate the situation. They first contacted the publisher of the other journal, Research India Publications (RIP), and also its Editor-in-Chief (EIC). The publisher did not respond, while the EIC reported that he himself was not able to contact the publisher! SIAM established that the whole paper published in IJSS, not just the abstract, was plagiarised verbatim. However, the plagiarised version was severely truncated, constituting just the last five pages of the 25-page SIAM paper, perhaps because RIP imposes a per page charge of $20 on the authors. Consequently, the paper published by RIP made little sense. It began abruptly without introduction, the abstract mostly summarised materials that had been removed from the paper, and the reference list mostly referred to such material.

The situation did not end there. SIAM found that the plagiarising authors had published or submitted at least half a dozen papers that were copied from other peoples' work. It then managed to get an admission from the authors by contacting their institutions with this evidence. Once the investigation was complete, SIAM wrote up the outcome in detail, posted it to a public web page, and informed many of the involved parties, including the authors of the original and copied articles and the editorial board of the RIP journal. At this point, eight months after our original inquiry, RIP finally responded to SIAM, with a one sentence email saying that it was removing the copied article from its website.

There is a clear financial aspect to these matters. RIP appears to be a for-profit operation, which publishes well over 100 journals in maths, science, and engineering. It sells subscriptions and collects authors' page charges. WorldCat lists 116 libraries that subscribe to the journal. Finally, we note that even a very clear-cut case, such as the one we outlined, could easily go undetected. Moreover, even when discovered, the sort of investigation described here is extremely time-consuming. More subtle ethical breaches, for example, plagiarism with paraphrasing, are correspondingly much more difficult to detect.

2 IMPACT FACTOR MANIPULATION

The impact factor is a bibliometric that has been widely adopted — by libraries, researchers, tenure and promotion committees, and editors and publishers — as a proxy for journal quality. It is calculated simply as the average number of citations in a year to the papers the journal published in the preceding two years. However, careful scrutiny shows that the impact factor is highly unreliable.

The impact factor for a journal in a given year is calculated by comparing the average number of citations in that year to the articles the journal published in the preceding two years. It has been widely criticised on a variety of grounds. Nonetheless, the lure of a simple number has proven irresistible to many, and consequently this measure has become a target for journal editors and publishers. Impact factor manipulation — gaming the system — has become common. Some journal editors pressure authors to add citations to boost the impact factor, essentially extorting journal citations in exchange for publication. Others publish review articles, which cite their journals profusely, or cultivate authors who can be counted upon for citations. Such practices distort the scientific literature, decreasing journal quality while artificially raising the impact factor.

Such manipulation is hard to detect and mostly known through anecdotal evidence. However, a study by librarian K. Fowler and the author into applied mathematics journals confirm that, indeed, the utility of the impact factor has been seriously compromised by manipulation. We scrutinised in detail the case of the International Journal of Nonlinear Sciences and Numerical Simulation (IJNSNS), which for the past four years has had the highest impact factor in the category of applied mathematics. Informed assessment of the journal is far different: IJNSNS did not even fall within the top 75 applied maths journals in a recent expert ranking by the Australian Research Council. Analysis of its anomalous impact factor revealed that a striking proportion of the citations to IJNSNS came from the editors of the journal itself. The journal's most prolific citer, by far, is its Editor-in-Chief; its second and third most regular citers are members of the editorial board as well. Together these three provide about a third of the citations year after year. Another third come from articles published in IJNSNS itself or in other journals and conference proceedings edited by the IJNSNS Editor-in-Chief. These numbers raise red flags, especially since similar checks of benchmark applied mathematics journals reveal only a very small proportion

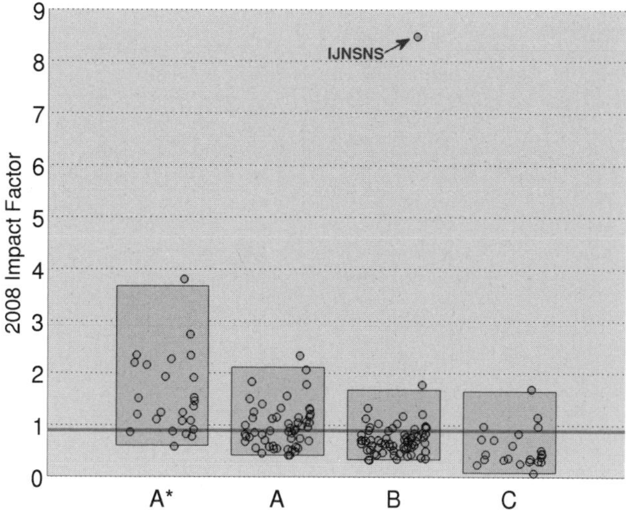

Fig. 1. 2008 impact factors of 170 applied math journals grouped according to their 2010 Excellence in Research for Australia rating tier. In each tier, the band runs from the 2.5th to the 97.5th percentile, outlining the middle 95%. Horizontal position of the data points within tiers is assigned randomly to improve visibility. The red line is at the 20th percentile of the A* tier.

of editor-connected citations. In further evidence of gaming, over 70% of citations to IJNSNS are to articles published in the previous two years, precisely the years that count toward the impact factor (in contrast to 10% which is typical for maths journals). Although IJNSNS is an extreme case, further investigation showed frequent discrepancies between expert judgment on journal quality and the impact factor. See Figure 1 for more details.

The continued use of the highly flawed impact factor has alarming consequences. Rewards are wrongly distributed, the scientific literature and enterprise are distorted, and cynicism about them grows. The international mathematics community is taking steps to address this challenge. As a start, the International Mathematical Union and the International Council for Industrial and Applied Mathematics recently voted to devise a system of rating mathematical journals based on expert judgment, not simple counting. A successful implementation of such a system would offer a sound journal assessment tool for librarians, editors, researchers, and others, and could provide a model to other scientific fields.

CHAPTER 48

PLAGIARISM UNDERSTANDING AND MANAGEMENT IN RUSSIA AND CENTRAL EUROPE

Vladimir Vlassov

The problem of plagiarism is the main focus of mass media discussions in music and literature in Eastern Europe, as well as in the West, but it has not been addressed openly in science or well-managed. For example, MEDLINE have only two records for plagiar* in Russian or from USSR/Russia since 1964. There has not been a single case of plagiarism detected in Russian medical periodicals for the last 20 years.

The detection of the plagiarism improved significantly in Russia between 2005 and 2008, after detection tools and search engines were employed by the best university teachers in Russia. This brought a better understanding of plagiarism to students, who were lucky to be taught by these teachers. As might be expected, students quickly start to use this technology to detect misconduct by their professors.

The understanding of plagiarism in society, as well as by scientists, is limited everywhere. Plagiarism of ideas is recognised, but is difficult to prove and is often controversial. Almost every time a Nobel prize is awarded, some scientists declare that their earlier works have a precedence and are not referenced in the laureate's publications. Because it is difficult to prove that these earlier publications influenced a laureate's work, the euphemism was introduced — "citation amnesia".

Many people limit plagiarism to exact copying of the text. Journal editors tend to pay attention only to this form of plagiarism. In Russia, some people even tend to excuse the exact copying as long as there is a reference to the source, regardless of the amount copied or absence of the quotation marks. One striking example of this misunderstanding occurred in 2010, when Russian Academy of Sciences (RAS) presidential commission looked at accusations in plagiarism against Vice-President A. Nekipelov. Nekipelov published a monograph that included a number of pages completely copied from other books. The commission concluded that because the plagiarised text was referenced, there was no plagiarism.

These problems are made worse in Russia by the continuing dominance of the old elite with powerful networking and group support. This is exemplified by the Kurjak case (Croatia). The "science elite" is connected to business and political "elites". Politicians and business people tend to excuse the research misconduct as a negligible problem. Only public pressure will change this situation. While J. Baiden (USA) plagiarized in presidential campaign, and ought to retire from campaign when he was caught, Vladimir Putin (Russia) plagiarized a candidate of science thesis, and never addressed the issue.

There is a tradition in Russia and some other post-Soviet countries of giving scientific degrees (candidate of science, doctor of science) to people who have had no "postgraduate" training but did somehow write a dissertation. This practice allows for people who really did no research to receive a degree. The proportion of these doctorates may be as large as one in three, as estimated by the Russian qualification body. It appears that the preservation of the tradition of awarding such degrees is a major barrier to setting higher standards for integrity in research and scientific publications. Russian academies and funding bodies urgently needed to establish committees to study research misconduct cases.

CHAPTER 49

BACKGROUND TO RESPONSIBLE RESEARCH PUBLICATION POSITION STATEMENTS[1]

Elizabeth Wager and Sabine Kleinert

The following position statements were developed at the 2nd World Conference on Research Integrity, held in Singapore in July 2010. They are designed to complement the Singapore Statement and to provide more detailed guidance on responsible research publication, with particular emphasis on research integrity and publication ethics. The first statement is aimed at researchers in their role as authors of publications. The second statement is aimed at editors of scholarly journals that publish research.

The two statements were originally drafted by the named authors (Elizabeth Wager and Sabine Kleinert, the Chair and Vice-Chair of the Committee on Publication Ethics (COPE)). These drafts were circulated before the meeting, discussed with the invited speakers, and revised to reflect these discussions. At the meeting in Singapore, the revised draft documents were presented and discussed in two sessions and further refined during a one-day, post-conference workshop. Both statements were then reworked to reflect the discussions in Singapore and circulated to those who had participated in the sessions and to members of the COPE Council and the International Council for Science (ICSU). However, while we hope that such organisations may endorse the statements, they are primarily based on the views of participants at the Singapore meeting and therefore

[1]contact details: liz@sideview.demon.co.uk, sabine.kleinert@lancet.com

do not necessarily represent the official views of any of the participating organisations or the individuals' institutions.

While some differences in publishing conventions exist between fields, it was evident from the discussion that there is much common ground, and also a desire to raise standards in the reporting of research. The two documents therefore aim to establish standards for authors and editors of scholarly research publications, and to describe responsible research reporting and publishing practice. Given the special issues raised by research involving humans or animals, which may not apply to other types of research, both statements include a specific section on these. We hope the statements will be endorsed by research institutions, funders, professional societies, and publishers.

While it would be impossible to reflect the views of all researchers and editors, we were pleased to involve participants and reviewers from a wide range of academic fields including biology, forestry, earth sciences, the humanities, mathematics, medicine, philosophy, and political science. The Singapore meeting also brought together participants from Africa, Asia, Australasia, Europe, the Middle East, and North America. We hope the versions presented here reflect the lively debate that took place before, during and after the meeting. However, we also hope that participants and reviewers will appreciate that it was not possible to incorporate all the suggestions we received, because some were contradictory. We therefore offer these documents as the first step in a process aimed at improving the reporting and publication of research and hope the statements will be reviewed and revised, as necessary, at future meetings.

We thank the following people who contributed to the discussions in Singapore and commented on drafts:

Siti Akmar Abu Samah (Universiti Teknologi MARA, Malaysia), Riaz Agha (*International Journal of Surgery*), Douglas Arnold (University of Minnesota and Society for Industrial and Applied Mathematics), Virginia Barbour (*Public Library of Science, PLoS Medicine*), Trish Groves (*BMJ*), Sara Jordan (Department of Politics and Public Administration, University of Hong Kong), Kamaruzaman Jusoff (Faculty of Forestry, Universiti Putra Malaysia), Abdellatif Maamri (Training Institute for Health Careers, Health Ministry, Oujda, Morocco), Ben Martin (*Research Policy*), Ana Marusic (*Croatian Medical Journal*), Linda Miller (*Nature*, now at New York University School of Medicine), Syntia Nchangwi (Cameroon), BJC Perera (*Sri Lanka Journal of Child Health, Sri Lanka Journal of Bio-Medical Informatics, Ceylon Medical Journal)*, Bernd Pulverer (European

Molecular Biology Organisation), Margaret Rees (*Maturitas*), Iveta Simera (EQUATOR Network), Randell Stephenson (*Journal of Geodynamics*), Xiongyong Sun (China National Knowledge Infrastructure), Diane Sullenberger (*Proceedings of the National Academy of Sciences*), David Vaux (La Trobe University, Australia), Vasiliy Vlassov (Society for Evidence Based Medicine, Moscow, Russia).

CHAPTER 50

RESPONSIBLE RESEARCH PUBLICATION: INTERNATIONAL STANDARDS FOR AUTHORS

A position statement developed at the 2nd World Conference on Research Integrity, Singapore, July 22–24, 2010

Elizabeth Wager and Sabine Kleinert[1]

- The research being reported should have been conducted in an ethical and responsible manner and should comply with all relevant legislation.
- Researchers should present their results clearly, honestly, and without fabrication, falsification or inappropriate data manipulation.
- Researchers should strive to describe their methods clearly and unambiguously so that their findings can be confirmed by others.
- Researchers should adhere to publication requirements that submitted work is original, is not plagiarised, and has not been published elsewhere.
- Authors should take collective responsibility for submitted and published work.
- The authorship of research publications should accurately reflect individuals' contributions to the work and its reporting.
- Funding sources and relevant conflicts of interest should be disclosed.

[1]liz@sideview.demon.co.uk, sabine.kleinert@lancet.com

INTRODUCTION

Publication is the final stage of research and therefore a responsibility for all researchers. Scholarly publications are expected to provide a detailed and permanent record of research. Because publications form the basis for both new research and the application of findings, they can affect not only the research community but also, indirectly, society at large. Researchers therefore have a responsibility to ensure that their publications are honest, clear, accurate, complete and balanced, and should avoid misleading, selective or ambiguous reporting. Journal editors also have responsibilities of ensuring the integrity of the research literature, and these are set out in companion guidelines.

This document aims to establish international standards for authors of scholarly research publications and to describe responsible research reporting practice. We hope these standards will be endorsed by research institutions, funders, and professional societies; promoted by editors and publishers; and will aid in research integrity training.

1 RESPONSIBLE RESEARCH PUBLICATION

1.1 *Soundness and Reliability*

1.1 The research being reported should have been conducted in an ethical and responsible manner and follow all relevant legislation. [*See also the Singapore Statement on Research Integrity, www.singaporestatement.org*]

1.2 The research being reported should be sound and carefully executed.

1.3 Researchers should use appropriate methods of data analysis and display (and, if needed, seek and follow specialist advice on this).

1.4 Authors should take collective responsibility for their work and for the content of their publications. Researchers should check their publications carefully at all stages to ensure methods and findings are reported accurately. Authors should carefully check calculations, data presentations, typescripts/submissions and proofs.

1.2 *Honesty*

2.1 Researchers should present their results honestly and without fabrication, falsification or inappropriate data manipulation. Research images (e.g. micrographs, X-rays, pictures of electrophoresis gels) should not be modified in a misleading way.

2.2 Researchers should strive to describe their methods and to present their findings clearly and unambiguously. Researchers should follow applicable reporting guidelines. Publications should provide sufficient detail to permit experiments to be repeated by other researchers.

2.3 Reports of research should be complete. They should not omit inconvenient, inconsistent or inexplicable findings or results that do not support the authors' or sponsors' hypothesis or interpretation.

2.4 Research funders and sponsors should not be able to veto the publication of findings that do not favour their product or position. Researchers should not enter agreements that permit the research sponsor to veto or control the publication of the findings (unless there are exceptional circumstances, such as research classified by governments because of security implications).

2.5 Authors should alert the editor promptly if they discover an error in any submitted, accepted or published work. Authors should cooperate with editors in issuing corrections or retractions when required.

2.6 Authors should represent the work of others accurately in citations and quotations.

2.7 Authors should not copy references from other publications if they have not read the cited work.

1.3 *Balance*

3.1 New findings should be presented in the context of previous research. The work of others should be fairly represented. Scholarly reviews and syntheses of existing research should be complete, balanced, and should include findings regardless of whether they support the hypothesis or interpretation being proposed. Editorials or opinion pieces presenting a single viewpoint or argument should be clearly distinguished from scholarly reviews.

3.2 Study limitations should be addressed in publications.

1.4 *Originality*

4.1 Authors should adhere to publication requirements that submitted work is original and has not been published elsewhere in any language. Work should not be submitted concurrently to more than one publication unless the editors have agreed to co-publication. If articles are co-published this fact should be made clear to readers.

4.2 Applicable copyright laws and conventions should be followed. Copyright material (e.g. tables, figures or extensive quotations) should be reproduced only with appropriate permission and acknowledgement.

4.3 Relevant previous work and publications, both by other researchers and the authors' own, should be properly acknowledged and referenced. The primary literature should be cited where possible.

4.4 Data, text, figures or ideas originated by other researchers should be properly acknowledged and should not be presented as if they were the authors' own. Original wording taken directly from publications by other researchers should appear in quotation marks with the appropriate citations.

4.5 Authors should inform editors if findings have been published previously or if multiple reports or multiple analyses of a single data set are under consideration for publication elsewhere. Authors should provide copies of related publications or work submitted to other journals.

4.6 Multiple publications arising from a single research project should be clearly identified as such and the primary publication should be referenced. Translations and adaptations for different audiences should be clearly identified as such, should acknowledge the original source, and should respect relevant copyright conventions and permission requirements. If in doubt, authors should seek permission from the original publisher before republishing any work.

1.5 *Transparency*

5.1 All sources of research funding, including direct and indirect financial support, supply of equipment or materials, and other support (such as specialist statistical or writing assistance) should be disclosed.

5.2 Authors should disclose the role of the research funder(s) or sponsor (if any) in the research design, execution, analysis, interpretation and reporting.

5.3 Authors should disclose relevant financial and non-financial interests and relationships that might be considered likely to affect the interpretation of their findings or which editors, reviewers or readers might reasonably wish to know. This includes any relationship to the journal, for example, if editors publish their own research in their own journal. In addition, authors should follow journal and institutional requirements for disclosing competing interests.

1.6 *Appropriate Authorship and Acknowledgement*

6.1 The research literature serves as a record not only of what has been discovered but also of who made the discovery. The authorship of research publications should therefore accurately reflect individuals' contributions to the work and its reporting.

6.2 In cases where major contributors are listed as authors while those who made less substantial, or purely technical, contributions to the research or to the publication are listed in an acknowledgement section, the criteria for authorship and acknowledgement should be agreed at the start of the project. Ideally, authorship criteria within a particular field should be agreed, published and consistently applied by research institutions, professional and academic societies, and funders. While journal editors should publish and promote accepted authorship criteria appropriate to their field, they cannot be expected to adjudicate in authorship disputes. Responsibility for the correct attribution of authorship lies with authors themselves working under the guidance of their institution. Research institutions should promote and uphold fair and accepted standards of authorship and acknowledgement. When required, institutions should adjudicate in authorship disputes and should ensure that due process is followed.

6.3 Researchers should ensure that only those individuals who meet authorship criteria (i.e. made a substantial contribution to the work) are rewarded with authorship and that deserving authors are not omitted. Institutions and journal editors should encourage practices that prevent guest, gift, and ghost authorship.

> guest authors are those who do not meet accepted authorship criteria but are listed because of their seniority, reputation or supposed influence.
> gift authors are those who do not meet accepted authorship criteria but are listed as a personal favour or in return for payment.
> ghost authors are those who meet authorship criteria but are not listed.

6.4 All authors should agree to be listed and should approve the submitted and accepted versions of the publication. Any change to the author list should be approved by all authors including any who have been removed from the list. The corresponding author should act as a point of contact between the editor and the other authors and should keep co-authors informed and involve them in major decisions about the publication (e.g. responding to reviewers' comments).

6.5 Authors should not use acknowledgements misleadingly to imply a contribution or endorsement by individuals who have not, in fact, been involved with the work or given an endorsement.

1.7 *Accountability and Responsibility*

7.1 All authors should have read and be familiar with the reported work and should ensure that publications follow the principles set out in these guidelines. In most cases, authors will be expected to take joint responsibility for the integrity of the research and its reporting. However, if authors take responsibility only for certain aspects of the research and its reporting, this should be specified in the publication.

7.2 Authors should work with the editor or publisher to correct their work promptly if errors or omissions are discovered after publication.

7.3 Authors should abide by relevant conventions, requirements, and regulations to make materials, reagents, software or datasets available to other researchers who request them. Researchers, institutions, and funders should have clear policies for handling such requests. Authors must also follow relevant journal standards. While proper acknowledgement is expected, researchers should not demand authorship as a condition for sharing materials.

7.4 Authors should respond appropriately to post-publication comments and published correspondence. They should attempt to answer correspondents' questions and supply clarification or additional details where needed.

1.8 *Adherence to Peer Review and Publication Conventions*

8.1 Authors should follow publishers' requirements that work is not submitted to more than one publication for consideration at the same time.

8.2 Authors should inform the editor if they withdraw their work from review, or choose not to respond to reviewer comments after receiving a conditional acceptance.

8.3 Authors should respond to reviewers' comments in a professional and timely manner.

8.4 Authors should respect publishers' requests for press embargoes and should not generally allow their findings to be reported in the press if they have been accepted for publication (but not yet published) in a scholarly publication. Authors and their institutions should liaise and cooperate with publishers to coordinate media activity (e.g. press releases and press conferences) around publication. Press releases should accurately reflect the work and should not include statements that go further than the research findings.

1.9 *Responsible Reporting of Research Involving Humans or Animals*

9.1 Appropriate approval, licensing or registration should be obtained before the research begins and details should be provided in the report (e.g. Institutional Review Board, Research Ethics Committee approval, national licensing authorities for the use of animals).

9.2 If requested by editors, authors should supply evidence that reported research received the appropriate approval and was carried out ethically (e.g. copies of approvals, licences, participant consent forms).

9.3 Researchers should not generally publish or share identifiable individual data collected in the course of research without specific consent from the individual (or their representative). Researchers should remember that many scholarly journals are now freely available on the internet, and should therefore be mindful of the risk of causing danger or upset to unintended readers (e.g. research participants or their families who recognise themselves from case studies, descriptions, images or pedigrees).

9.4 The appropriate statistical analyses should be determined at the start of the study and a data analysis plan for the prespecified outcomes should be prepared and followed. Secondary or *post hoc* analyses should be distinguished from primary analyses and those set out in the data analysis plan.

9.5 Researchers should publish all meaningful research results that might contribute to understanding. In particular, there is an ethical responsibility to publish the findings of all clinical trials. The publication of unsuccessful studies or experiments that reject a hypothesis may help prevent others from wasting time and resources on similar projects. If findings from small studies and those that fail to reach statistically significant results can be combined to produce more useful information (e.g. by meta-analysis) then such findings should be published.

9.6 Authors should supply research protocols to journal editors if requested (e.g. for clinical trials) so that reviewers and editors can compare the research report to the protocol to check that it was carried out as planned and that no relevant details have been omitted. Researchers should follow relevant requirements for clinical trial registration and should include the trial registration number in all publications arising from the trial.

CHAPTER 51

RESPONSIBLE RESEARCH PUBLICATION: INTERNATIONAL STANDARDS FOR EDITORS

A position statement developed at the 2nd World Conference on Research Integrity, Singapore, July 22–24, 2010

Sabine Kleinert and Elizabeth Wager[1]

SUMMARY

- Editors are accountable and should take responsibility for everything they publish.
- Editors should make fair and unbiased decisions independent from commercial consideration and ensure a fair and appropriate peer review process.
- Editors should adopt editorial policies that encourage maximum transparency and complete, honest reporting.
- Editors should guard the integrity of the published record by issuing corrections and retractions when needed, and pursuing suspected or alleged research and publication misconduct.
- Editors should pursue reviewer and editorial misconduct.
- Editors should critically assess the ethical conduct of studies in humans and animals.
- Peer reviewers and authors should be told what is expected of them.
- Editors should have appropriate policies in place for handling editorial conflicts of interest.

[1]sabine.kleinert@lancet.com, liz@sideview.demon.co.uk

INTRODUCTION

As guardians and stewards of the research record, editors should encourage authors to strive for, and adhere themselves to, the highest standards of publication ethics. Furthermore, editors are in a unique position to indirectly foster responsible conduct of research through their policies and processes. To achieve the maximum effect within the research community, ideally all editors should adhere to universal standards and good practices. While there are important differences between different fields and not all areas covered are relevant to each research community, there are important common editorial policies, processes, and principles that editors should follow to ensure the integrity of the research record.

These guidelines are a starting point and are aimed at journal editors in particular. While books and monographs are important and relevant research records in many fields, guidelines for book editors are beyond the scope of these recommendations. It is hoped that in due course such guidelines can be added to this document.

Editors should regard themselves as part of the wider professional editorial community, keep themselves abreast of relevant policies and developments, and ensure their editorial staff is trained and kept informed of relevant issues.

To be a good editor requires many more principles than are covered here. These suggested principles, policies, and processes are particularly aimed at fostering research and publication integrity.

1 EDITORIAL PRINCIPLES

1.1 *Accountability and Responsibility for Journal Content*

Editors have to take responsibility for everything they publish and should have procedures and policies in place to ensure the quality of the material they publish and maintain the integrity of the published record (see paragraphs 4–8).

1.2 *Editorial Independence and Integrity*

An important part of the responsibility to make fair and unbiased decisions is the upholding of the principle of editorial independence and integrity.

1.2.1 *Separating Decision-Making from Commercial Considerations*

Editors should make decisions on academic merit alone and take full responsibility for their decisions. Processes must be in place to separate commercial activities within a journal from editorial processes and decisions. Editors should take an active interest in the publisher's pricing policies and strive for wide and affordable accessibility of the material they publish.

Sponsored supplements must undergo the same rigorous quality control and peer review as any other content for the journal. Decisions on such material must be made in the same way as any other journal content. The sponsorship and role of the sponsor must be clearly declared to readers.

Advertisements need to be checked so that they follow journal guidelines, should be clearly distinguishable from other content, and should not in any way be linked to scholarly content.

1.2.2 *Editors' Relationship to the Journal Publisher or Owner*

Editors should ideally have a written contract setting out the terms and conditions of their appointment with the journal publisher or owner. The principle of editorial independence should be clearly stated in this contract. Journal publishers and owners should not have any role in decisions on content for commercial or political reasons. Publishers should not dismiss an editor because of any journal content unless there was gross editorial misconduct or an independent investigation has concluded that the editor's decision to publish was against the journal's scholarly mission.

1.2.3 *Journal Metrics and Decision-Making*

Editors should not attempt to inappropriately influence their journal's ranking by artificially increasing any journal metric. For example, it is inappropriate to demand that references to that journal's articles are included except for genuine scholarly reasons. In general, editors should ensure that papers are reviewed on purely scholarly grounds and that authors are not pressured to cite specific publications for non- scholarly reasons.

1.3 *Editorial Confidentiality*

1.3.1 *Authors' Material*

If a journal operates a system where peer reviewers are chosen by editors (rather than posting papers for all to comment as a pre-print version), editors must protect the confidentiality of authors' material and remind reviewers to do so as well. In general, editors should not share submitted papers with editors of other journals, unless with the authors' agreement or in cases of alleged misconduct (see below). Editors are generally under no obligation to provide material to lawyers for court cases. Editors should not give any indication of a paper's status with the journal to anyone other than the authors. Web-based submission systems must be run in a way that prevents unauthorised access.

In the case of a misconduct investigation, it may be necessary to disclose material to third parties (e.g. an institutional investigation committee or other editors).

1.3.2 *Reviewers*

Editors should protect reviewers' identities unless operating an open peer review system. However, if reviewers wish to disclose their names, this should be permitted. If there is alleged or suspected reviewer misconduct, it may be necessary to disclose a reviewer's name to a third party.

2 GENERAL EDITORIAL POLICIES

2.1 *Encourage Maximum Transparency and Complete and Honest Reporting*

To advance knowledge in scholarly fields, it is important to understand why particular work was done, how it was planned and conducted and by whom, and what it adds to current knowledge. To achieve this understanding, maximum transparency and complete and honest reporting are crucial.

2.1.1 *Authorship and Responsibility*

Journals should have a clear policy on authorship that follows the standards within the relevant field. They should give guidance in their information for authors on what is expected of an author and, if there are different authorship conventions within a field, they should state which they adhere to.

For multidisciplinary and collaborative research, it should be apparent to readers who has done what and who takes responsibility for the conduct and validity of which aspect of the research. Each part of the work should have at least one author who takes responsibility for its validity. For example, individual contributions and responsibilities could be stated in a contributor section. All authors are expected to have contributed significantly to the paper and to be familiar with its entire content and ideally, this should be declared in an authorship statement submitted to the journal.

When there are undisputed changes in authorship for appropriate reasons, editors should require that all authors (including any whose names are being removed from an author list) agree to these in writing. Authorship disputes (i.e. disagreements on who should or should not be an author before or after publication) cannot be adjudicated by editors and should be resolved at institutional level or through other appropriate independent bodies for both published and unpublished papers. Editors should then act on the findings, for example, by correcting authorship in published papers. Journals should have a publicly declared policy on how papers submitted by editors or editorial board members are handled (see paragraph on editorial conflicts of interest: 4.2.2).

2.1.2 *Conflicts of Interest and Role of the Funding Source*

Editors should have policies that require all authors to declare any relevant financial and non-financial conflicts of interest and publish at least those that might influence a reader's perception of a paper alongside the paper. The funding source of the research should be declared and published, and the role of the funding source in the conception, conduct, analysis, and reporting of the research should be stated and published.

Editors should make it clear in their information for authors if, in certain sections of the journal (e.g. commissioned commentaries or review articles), certain conflicts of interest preclude authorship.

2.1.3 *Full and Honest Reporting and Adherence to Reporting Guidelines*

Among the most important responsibilities of editors is to maintain a high standard in the scholarly literature. Although standards differ among journals, editors should work to ensure that all published papers make a substantial new contribution to their field. Editors should discourage so-called "salami publications" (i.e. publication of the minimum publishable unit of research), avoid duplicate or redundant publication unless it is fully declared and acceptable to all (e.g. publication in a different language with cross-referencing), and encourage authors to place their work in the context of previous work (i.e. to state why this work was necessary/done, what this work adds or why a replication of previous work was required, and what readers should take away from it).

Journals should adopt policies that encourage full and honest reporting, for example, by requiring authors in fields where it is standard to submit protocols or study plans, and, where they exist, to provide evidence of adherence to relevant reporting guidelines. Although devised to improve reporting, adherence to reporting guidelines also makes it easier for editors, reviewers, and readers to judge the actual conduct of the research.

Digital image files, figures, and tables should adhere to the appropriate standards in the field. Images should not be inappropriately altered from the original or present findings in a misleading way.

Editors might also consider screening for plagiarism, duplicate or redundant publication by using anti-plagiarism software, or for image manipulation. If plagiarism or fraudulent image manipulation is detected, this should be pursued with the authors and relevant institutions (see paragraph on how to handle misconduct: 2.2.2)

2.2 *Responding to Criticisms and Concerns*

Reaction and response to published research by other researchers is an important part of scholarly debate in most fields and should generally be encouraged. In some fields, journals can facilitate this debate by publishing readers' responses. Criticisms may be part of a general scholarly debate but can also highlight transgressions of research or publication integrity.

2.2.1 *Ensuring Integrity of the Published Record — Corrections*

When genuine errors in published work are pointed out by readers, authors, or editors, which do not render the work invalid, a correction (or erratum) should be published as soon as possible. The online version of the paper may be corrected with a date of correction and a link to the printed erratum. If the error renders the work or substantial parts of it invalid, the paper should be retracted with an explanation as to the reason for retraction (i.e. honest error).

2.2.2 *Ensuring the Integrity of the Published Record — Suspected Research or Publication Misconduct*

If serious concerns are raised by readers, reviewers, or others, about the conduct, validity, or reporting of academic work, editors should initially contact the authors (ideally all authors) and allow them to respond to the concerns. If that response is unsatisfactory, editors should take the concerns to the institutional level (see below). In rare cases, mostly in the biomedical field, when concerns are very serious and the published work is likely to influence clinical practice or public health, editors should consider informing readers about these concerns, for example by issuing an "expression of concern", while the investigation is ongoing. Once an investigation is concluded, the appropriate action needs to be taken by editors with an accompanying comment that explains the findings of the investigation. Editors should also respond to findings from national research integrity organisations that indicate misconduct relating to a paper published in their journal. Editors can themselves decide to retract a paper if they are convinced that serious misconduct has happened even if an investigation by an institution or national body does not recommend it.

Editors should respond to all allegations or suspicions of research or publication misconduct raised by readers, reviewers, or other editors. Editors are often the first recipients of information about such concerns and should act, even in the case of a paper that has not been accepted or has already been rejected. Beyond the specific responsibility for their journal's publications, editors have a collective responsibility for the research record and should act whenever they become aware of potential misconduct if at all possible. Cases of possible plagiarism or duplicate/redundant publication can be assessed by editors themselves. However, in most other cases, editors should request an investigation by the institution or other appropriate bodies (after seeking an explanation from the authors first and if that explanation is unsatisfactory).

Retracted papers should be retained online, and they should be prominently marked as a retraction in all online versions, including the PDF, for the benefit of future readers. For further guidance on specific allegations and suggested actions, such as retractions, see the COPE flowcharts and retraction guidelines (http://publicationethics.org/flowcharts; http://publicationethics.org/files/u661/Retractions_COPE_gline_final_3_Sept_09_2_.pdf).

2.2.3 *Encourage Scholarly Debate*

All journals should consider the best mechanism by which readers can discuss papers, voice criticisms, and add to the debate (in many fields this is done via a print or online correspondence section). Authors may contribute to the debate by being allowed to respond to comments and criticisms where relevant. Such scholarly debate about published work should happen in a timely manner. Editors should clearly distinguish between criticisms of the limitations of a study and criticisms that raise the possibility of research misconduct. Any criticisms that raise the possibility of misconduct should not just be published but should be further investigated even if they are received a long time after publication.

3 EDITORIAL POLICIES RELEVANT ONLY TO JOURNALS THAT PUBLISH RESEARCH IN HUMANS OR ANIMALS

3.1 *Critically Assess and Require a High Standard of Ethical Conduct of Research*

Especially in biomedical research but also in social sciences and humanities, ethical conduct of research is paramount in the protection of humans and animals. Ethical oversight, appropriate consent procedures, and adherence to relevant laws are required from authors. Editors need to be vigilant to concerns in this area.

3.1.1 *Ethics Approval and Ethical Conduct*

Editors should generally require approval of a study by an ethics committee (or institutional review board) and the assurance that it was conducted according to the Declaration of Helsinki for medical research in humans but, in addition, should be alert to areas of concern in the ethical conduct of research. This may mean that a paper is sent to peer reviewers who possess particular expertise in this area, to the journal's ethics committee if there is one, or that editors require further reassurances or evidence from authors or their institutions.

Papers may be rejected on ethical grounds even if the research had ethics committee approval.

3.1.2 *Consent (to Take Part in Research)*

If research is done on humans, editors should ensure that a statement on the consent procedure is included in the paper. In most cases, written informed consent is the required norm. If there is any concern about the consent procedure, if the research is done in vulnerable groups, or if there are doubts about the ethical conduct, editors should ask to see the consent form and enquire further from authors as to exactly how consent was obtained.

3.1.3 *Consent (for Publication)*

For all case reports, small case series, and images of people, editors should require the authors to have obtained explicit consent for publication (which is different from consent to take part in research). This consent should inform participants which journal the work will be published in, make it clear that, although all efforts will be made to remove unnecessary identifiers, complete anonymity is not possible, and ideally state that the person described has seen and agreed with the submitted paper. The signed consent form should be kept with the patient file rather than be sent to the journal (to maximise data protection and confidentiality, see paragraph 6.4). There may be exceptions where it is not possible to obtain consent, for example when the person has died. In such cases, a careful consideration about possible harm is needed and, out of courtesy, attempts should be made to obtain assent from relatives. In very rare cases, an important public health message may justify publication without consent if it is not possible, despite all efforts, to obtain consent, and the benefit of publication outweighs the possible harm.

3.1.4 *Data Protection and Confidentiality*

Editors should critically assess any potential breaches of data protection and patient confidentiality. This includes requiring properly informed consent for the actual research presented, consent for publication where applicable (see paragraph 6.3), and having editorial policies that comply with guidelines on patient confidentiality.

3.1.5 *Adherence to Relevant Laws and Best Practice Guidelines for Ethical Conduct*

Editors should require authors to adhere to relevant national and international laws and best practice guidelines where applicable, for example when undertaking animal research. Editors should encourage registration of clinical trials.

4 EDITORIAL PROCESSES

4.1 *Ensuring a Fair and Appropriate Peer Review Process*

One of the most important responsibilities of editors is organising and using peer review fairly and wisely. Editors should explain their peer review processes in the information for authors and also indicate which parts of the journal are peer reviewed.

4.1.1 *Decision Whether to Review*

Editors may reject a paper without peer review when it is deemed unsuitable for the journal's readers or is of poor quality. This decision should be made in a fair and unbiased way. The criteria used to make this decision should be made explicit. The decision not to send a paper for peer review should only be based on the academic content of the paper, and should not be influenced by the nature of the authors or the host institution.

4.1.2 *Interaction with Peer Reviewers*

Editors should use appropriate peer reviewers for papers that are considered for publication by selecting people with sufficient expertise and avoiding those with conflicts of interest. Editors should ensure that reviews are received in a timely manner.

Peer reviewers should be told what is expected of them and should be informed about any changes in editorial policies. In particular, peer reviewers should be asked to assess research and publication ethics issues (i.e. whether they think the research was done and reported ethically, or if they have any suspicions of plagiarism, fabrication, falsification, or redundant publication). Editors should have a policy to request a formal conflict of interest declaration from peer reviewers and should ask peer reviewers to inform them about any such conflict of interest at the earliest opportunity so that they can make a decision on whether an unbiased review is possible.

Certain conflicts of interest may disqualify a peer reviewer. Editors should stress confidentiality of the material to peer reviewers and should require peer reviewers to inform them when they ask a colleague for help with a review or if they mentor a more junior colleague in conducting peer review. Editors should ideally have a mechanism to monitor the quality and timeliness of peer review and to provide feedback to reviewers.

4.1.3 *Reviewer Misconduct*

Editors must take reviewer misconduct seriously and pursue any allegation of breach of confidentiality, non-declaration of conflicts of interest (financial or non-financial), inappropriate use of confidential material, or delay of peer review for competitive advantage. Allegations of serious reviewer misconduct, such as plagiarism, should be taken to the institutional level (for further guidance see: http://publicationethics.org/files/u2/07_Reviewer_misconduct.pdf).

4.1.4 *Interaction with Authors*

Editors should make it clear to authors what the role of the peer reviewer is, because this may vary from journal to journal. Some editors regard peer reviewers as advisors and may not necessarily follow (or even ask for) reviewers' recommendations on acceptance or rejection. Correspondence from editors is usually with the corresponding author, who should guarantee the involvement of co-authors at all stages. Communicating with all authors at first submission and at final acceptance stage can be helpful to ensure all authors are aware of the submission and have approved the publication. Normally, editors should pass on all peer reviewers' comments in their entirety. However, in exceptional cases, it may be necessary to exclude parts of a review, if it, for example, contains libellous or offensive remarks. It is important, however, that such editorial discretion is not inappropriately used to suppress inconvenient comments.

There should always be good reasons, which are clearly communicated to authors, if additional reviewers are sought at a late stage in the process.

The final editorial decision and reasons for this should be clearly communicated to authors and reviewers. If a paper is rejected, editors should ideally have an appeal process. Editors, however, are not obliged to overturn their decision.

4.2 Editorial Decision-Making

Editors are in a powerful position by making decisions on publications, which makes it very important that this process is as fair and unbiased as possible, and is in accordance with the academic vision of the particular journal.

4.2.1 Editorial and Journal Processes

All editorial processes should be made clear in the information for authors. In particular, it should be stated what is expected of authors, which types of papers are published, and how papers are handled by the journal. All editors should be fully familiar with the journal policies, vision, and scope. The final responsibility for all decisions rests with the editor-in-chief.

4.2.2 Editorial Conflicts of Interest

Editors should not be involved in decisions about papers in which they have a conflict of interest, for example, if they work or have worked in the same institution and collaborated with the authors, if they own stock in a particular company, or if they have a personal relationship with the authors. Journals should have a defined process for handling such papers. Journals should also have a process in place to handle papers submitted by editors or editorial board members to ensure unbiased and independent handling of such papers. This process should be stated in the information for authors. Editorial conflicts of interests should be declared, ideally publicly.

SECTION VIII

INTEGRITY IN THE NEWS, CLIMATE CHANGE AND DUAL-USE TECHNOLOGY

INTRODUCTION

The sessions devoted to "Research Integrity in the News" focused on two topical issues in which integrity and public interest coincide in a major way: Climate Change and Dual-use Technology.

The session on Climate Research took place in the shadow of the so-called "ClimateGate" affair at the University of East Anglia (UEA) in the UK and concerns about the integrity and validity of the latest reports of the Inter-Governmental Panel on Climate Change (IPCC) — a Nobel Prizewinner. It also follows in the aftermath of the failure of the World Climate Conference in Copenhagen in December 2009, which was influenced, at least to some degree, by the UEA and IPCC accusations.

In his contribution, Frankel clearly spells out the implications from these cases, of the problems of the interaction between research and society at large, transparency versus secrecy and the impression given frequently that the research world is closed to outsiders. He highlights the balance that needs to be struck between researchers when speaking as experts and when speaking as partisans for one cause or another. He also poses a number of questions that merit and require further debate and consideration:

- How can/does advocacy detract from the objectivity and dispassion typically expected of scientists?
- What are the implications of the public's need for reliable and independent advice on highly technical matters?
- When do scientists cross the line from being an independent source of valued information to designing or using their researches to support some preconceived policy preferences?
- What is meant by "responsible advocacy"?
- Are there "rules of the road" or best practices that can guide the scientist-advocate?
- How do we teach responsible advocacy?

In his contribution, Pearce speaks from the viewpoint of a responsible journalist who has followed the climate debate and, in particular, the UEA "ClimateGate". He demonstrates, through his account of the affair, the need for transparency coupled with the public obligations of researchers, especially in an area so close to policymaking at all levels. Henderson-Sellers reinforces the point that there may be a special conflict in climate science between the "meritocratic" processes of research and the "democratic" decision-making processes based heavily on that research. This again highlights need for the individual responsibility in public advocacy recommended in Responsibility 10 of the Singapore Statement.

Finally, it should be borne in mind that "Earth System Science" represents the ultimate in the globalisation of research. Within the mega-projects of the World Climate Research Programme (WCRP), the International Geosphere-Biosphere Programme (IGBP), the International Human Dimensions Programme (IHDP) and Diversitas, which together make up the Earth Science Systems Partnership (ESSP), there are numerous other specific thematic projects that are also of a global nature. Given the many individuals and research groups from many nations across the world involved in ESSP, the questions of quality assurance, especially the reliability of data and observations, and research integrity are vital. If anything were to be amiss in terms of misconduct or indeed serious questionable and negligent practices, it would cast doubt on this major research edifice. How to promote research integrity in these complex, global and multidisciplinary programmes is a significant challenge that has to be addressed by the mega-projects organisers, especially following the recent experiences in climate science discussed in this section. "ClimateGate" illustrated that it is very easy to lose public trust and very hard to regain it.

Dual-use is another area of research where public trust is a crucial factor, but this time the trust is not so much about the integrity of the research itself but its capacity for misuse. The issue considered in this session especially addressed how to combat misuse and the role of codes of conduct in awareness-raising within the research community.

For many, the term "dual-use" is difficult to understand. It basically refers to the use of research findings that could be both beneficial and harmful. Sometimes, this is the appropriation of "blue skies" research for military or security purposes, when originally not intended for this end, but it now also can include the use of research findings for terrorism. Anestidou, in her introductory paper, points out that recent attention in the dual-use debate has focused on life sciences research. She asks whether "we should forego the quest for truth ... for fear of its outcomes?" This is the age-old problem of how to put the genie back in the bottle once released. Anestidou points out that the researchers believe that these questions are best addressed by the research community through self-governance.

Epstein explores this subject further and points out that the life sciences are not unique in posing a dual-use threat. This was the issue decades ago with regard to nuclear research and its possible misuse. However, Epstein highlights the new problem that has arisen due to the nihilistic beliefs and actions of those bent on terrorism. As with Anestidou, Epstein points out the need for self-policing, which can be faster than legislation and regulation in setting standards through norms and codes of conduct.

Franz turns his attention to the role of leadership and the establishment of a culture of responsibility within laboratories, especially through personnel reliability and what he terms agent accountability. He points out that, generally, a culture that builds trust and transparency will be more productive and safer than one that is authoritarian and mistrusts its employees. However, we are considering areas of activity in which government oversight tends to be more intrusive because of the threats posed to society at large. Taking a line from safety issues in other contexts is suggested as one way forward, but Franz insists that enlightened leadership is one of the best safeguards against misconduct in this sensitive area.

Davis, in his contribution, outlines the history of dual-use codes of conduct through the US National Science Advisory Board for Biosecurity. Again, emphasis is laid on the responsibility of the individual researcher in the oversight of possible dual-use life sciences research. Scientists play the "pivotal role". This leads to Davis's question — "How is awareness of the dual-use dilemma to be promoted and fostered among scientists themselves?" Further work in promoting and adopting codes of conduct will be necessary. He then turns to "moral agency". In other words, if there is character deficiency or a lack of morality in the individual, then no norm or code of conduct can compensate for this. He concludes that codes do have the benefit of raising awareness and that this role is "potentially significant" in enhancing knowledge of the dilemma faced in dual-use research.

As in other aspects of research integrity, the foundation for good conduct is education. Heitman explores whether education about the responsible conduct of research is appropriate in the dual-use area. As she mentions, educators have often considered laboratory and environmental safety and biosafety as appropriate areas of teaching. The dual-use dilemma can be used to illustrate more general aspects of research integrity and especially in bringing ethical issues into consideration. It can also show that the impact of discovery is itself uncertain. We are all aware of this uncertainty in terms of the exploitation of research for innovation. Heitman discusses how what she terms "professional self governance" can address complex questions "beyond the reach of law". So dual-use, or the misuse of research for truly antisocial purposes, is a new ethical challenge that educators have to face. From the evidence from American universities, it seems that "honour codes" may have a more profound impact than the simple rigid application of codes, which may be more honoured in the breach than in the observance. She concludes that dual-use issues need to be addressed in the education of the coming generation of researchers.

Both sessions of the conference showed that research integrity is not an abstruse aspect of research, but has a direct bearing on research that is highly visible to society, and in which the maintenance of public trust is essential. The behaviour of researchers is ultimately a matter of individual responsibility, but we need to be aware, both as individuals and collectively, that we have to show responsibility and behave with integrity to retain public support for our endeavours.

CHAPTER 52

TURNING UP THE HEAT ON RESEARCH INTEGRITY: LESSONS FROM "CLIMATEGATE"

Mark S Frankel

"Climategate" refers to events that began in November 2009, with the internet leak of volumes of e-mail traffic among climate scientists involved with a global climate assessment prepared by the Intergovernmental Panel on Climate Change (IPCC). Release of the e-mails (and related documents) led to allegations of research misconduct against some of the world's leading climate scientists, who were accused of making global warming appear more serious than it is (Hickman and Randerson, 2009). Several subsequent independent inquiries found no evidence of misconduct, but some of the practices of the accused climate researchers were criticised (Adam, 2010).

While Climategate has left intact the strong consensus that global warming is occurring and that it is largely precipitated by human actions, it has nevertheless raised important issues for the larger scientific community that strike at the core of research integrity. This essay examines those issues in the context of the relationship between science and society, and more specifically the engagement of scientists in the public arena, where their research data and findings may be used in crafting and implementing public policy.

1 SCIENCE AND SOCIETY

Modern societies depend heavily on expert knowledge to solve pressing social problems. Yet people recognise that advances in science can also lead to unwelcome consequences, and they are increasingly aware that science, and scientists are influenced by powerful political, economic and social forces. This understanding takes on special significance when scientific expertise is used in formulating and implementing public policy, where the line between advice and advocacy is often blurred.

Although scientists' involvement in policy deliberations is not a new development, the proper role of scientists in the policy process has become particularly contentious in recent years. There are strong expectations by both scientists and the public that research will be conducted in a manner consistent with accepted norms and practices, and that its reporting by scientists will be a fair and honest representation of the research; in other words, that the research is trustworthy.

The connection to policy is reflected in a statement by US President Barack Obama in March 2009, "The public must be able to trust the science and scientific process informing public policy decisions." As a result, "Trust has been damaged," according to a leading climate researcher from Germany. "People now find it conceivable that scientists cheat and manipulate..." (quoted in Pearce, 2010). One lesson of Climategate that applies to research more broadly is that "it is no longer tenable to believe that... trusted scientific knowledge can come into existence inside laboratories that are hermetically sealed" from public scrutiny (Hulme and Ravetz, 2009). Research integrity, once thought to be the sole domain of scientists, is now subject to greater examination by an expanding number of stakeholders who perceive a relationship between research and their core concerns.

The loss of trust can be further compounded not only by what scientists may do, but also by how their actions are perceived by others. A recent editorial on the Climategate episode in the journal *Nature* (2010, p. 7) entitled "A Question of Trust", observes that "people — politicians included — make decisions on the basis of self-interest and their own hopes, fears and values, which will not necessarily match what many researchers deem self-evident." Note that terms such as "facts", "data", and "evidence", the bread and butter of scientists, do not appear in that observation. People are generally too far removed from debates over science and policy to truly understand the "facts" or potential consequences of the situation. As a result, peoples' actions will often be driven by how they perceive the situation relative to their self-interest and values (Edelman, 1964). In such circumstances, facts matter, but not necessarily because of the knowledge they presumably convey. What matters is how they are perceived by others.

The UK House of Commons Science and Technology Committee inquiry into the leaked emails got it right when it stated that, "Reputation does not, however, rest solely on the quality of work... It also depends on perception. It is self-evident that the disclosure of ... e-mails has damaged the reputation of UK climate science and, as views on global warming have become polarised, any deviation from the highest scientific standards will be pounced on" (House of Commons, 2010, p. 44).

2 SCIENTISTS AS EXPERTS IN THE POLICY ARENA

The role of scientists in the policy arena is of considerable importance relative to research integrity. It speaks directly to questions about objectivity, disinterestedness, and advocacy. Writing at the height of concerns about the behaviour of climate scientists, one commentator wrote that "professional objectivity is precisely what the hacked e-mails call into question. Some of these scientists are merely activists, deeply invested in a predetermined outcome" (Gerson, 2009). Another echoed this concern, writing that the e-mails "reveal... something problematic for the scientific community as a whole, namely, the tendency of scientists to cross the line from being disinterested investigators after the truth to advocates for a preconceived conclusion about the issues at hand" (Hayward, 2009).

With the possibility that advocacy can undermine scientific independence and credibility, both scientists and the public should consider the following:

- How can/does advocacy detract from the objectivity and dispassion typically expected of scientists? What are the implications for the public's need for reliable and independent advice on highly technical matters?
- When do scientists cross the line from being an independent source of valued information to designing or using their research to support some preconceived policy preference?
- What is meant by "responsible advocacy"? Are there "rules of the road" or best practices that can guide the scientist-advocate? How do we teach responsible advocacy?

References

1. Adam, D. (7 July 2010). *Climategate' Review Clears Scientists of Dishonesty Over Data.* Accessed from http://www.guardian.co.uk/environment/2010/jul/07/climategate-review-clears-scientists-dishonesty.

2. A Question of Trust. Editorial. *Nature* (1 July 2010). Accessed from http://www.nature.com/nature/journal/v466/n7302/full/466007a.html.
3. Edelman, M. (1964). *The Symbolic Uses of Politics.* IL: University of Illinois Press.
4. Gerson, M. (11 December 2009). *Whose War on Science?.* Accessed from http://www.washingtonpost.com/wp-dyn/content/article/2009/12/10/AR2009121003159.html.
5. Hayward, S. F. (14 December 2009). *Scientists Behaving Badly.* Accessed from http://www.weeklystandard.com/Content/Public/Articles/000/000/017/300ubchn.asp?pg=1.
6. Hickman, L. and Randerson, J. (20 November 2009). *Climate Sceptics Claim Leaked Emails are Evidence of Collusion among Scientists.* Accessed from http://www.guardian.co.uk/environment/2009/nov/20/climatesceptics-hackers-leaked-emails.
7. House of Commons, Science and Technology Committee (31 March 2010). *The disclosure of climate data from the Climatic Research Unit at the University of East Anglia.* Accessed from http://www.publications.parliament.uk/pa/cm200910/cmselect/cmsctech/387/38702.htm.
8. Hulme, M. and Ravetz, J. (1 December 2009). *Show Your Working: What 'Climate Gate' Means.* Accessed from http://news.bbc.co.uk/2/hi/science/nature/8388485.stm.
9. Obama, B. (9 March 2009). *Memorandum For The Heads Of Executive Departments And Agencies.* Accessed from http://www.whitehouse.gov/the_press_office/Memorandum-for-the-Heads-of-Executive-Departments-and-Agencies-3-9-09/.
10. Pearce, F. (4 July 2010). *'Climategate' was a 'game-changer' in science reporting, say climatologists.* Accessed from http://www.guardian.co.uk/environment/2010/jul/04/climatechange-hacked-emails-muir-russell.

CHAPTER 53

CLIMATEGATE: A JOURNALIST'S PERSPECTIVE

Fred Pearce

The public by and large believes in the integrity of scientists. That is why science is mostly allowed to police itself. Through peer review, assessment processes like the Intergovernmental Panel on Climate Change — where climate scientists review the work of other climate scientists — and so on.

But few professions are allowed this luxury today. And there are cracks in the edifice that suggest science may face much greater scrutiny in future.

Climate science in particular has taken some hits in the last year. This includes the notable saga that has become known as Climategate — the fallout from the release of a thousand e-mails held over 14 years on the web server of the Climatic Research Unit of the University of East Anglia in England.

While the three inquiries held in Britain into the affair have, partly at least, exonerated the scientists of the worst charges against them, questions do remain. And the US Congress is currently considering whether it should continue to fund CRU's key research into the thermometer records of climate over the past 160 years — the research at the centre of the e-mails row.

As a journalist specialising in climate science, I have written a lot about this affair. I wrote an investigation for the Guardian newspaper in England, which we turned into a book. This is the cover. I think it raises important questions.

The Climategate story "broke" one Friday lunchtime in late November last year. Editors were on the phone. E-mails from the University of East Anglia had been copied and published on the internet.

I was one of the few journalists to speak to Phil Jones, the director of CRU and one of the chief e-mailers before his university imposed a news blackout. It was a theft, the university said. There was a crime, end of story.

But the trouble was that the e-mails were out there. And bloggers, most of them climate sceptics, were spinning them as a smoking gun that revealed a global conspiracy by climate scientists to lie to us about climate change.

As I read the e-mails over that weekend, two things became clear. First, there was no smoking gun over the science. But second, there was an awful lot of rather unpleasant e-mail chats among CRU scientists and their correspondents in the USA about how to marginalise, bad-mouth and silence their many critics. Some of it looked like incitement to break the law. I was shocked.

It seemed to me that there was a war going on. This was partly, of course, a rhetorical battle for the agenda on climate change. But there was also a battle for ownership of climate data. At the root, the closed world of peer review was under challenge by people who were not so much climate sceptics as data libertarians.

There exist people like the Canadian statistician Steve McIntyre. He believes in climate change, but he also thinks he should have the right to do his own independent analysis of data collected by scientists like Jones.

During December, it became clear that the e-mails were political dynamite, and were damaging public confidence in the science of climate change. Phrases like "hide the decline" — three words out of a million in the e-mails — were having a real impact on perceptions about climate change across the world.

This propaganda was built on a lie. Leading US politicians like Sarah Palin, who may have misunderstood, and Senator James Inhoff of Oklahoma, who should have known better, told the world that the phrase showed climate scientists were manipulating data to "hide the decline in global temperatures" over the past decade.

Nonsense. How could it have been anything else, when the e-mail itself was written in 1999 before the decline (actually just a levelling off) had begun? It was written at the end of the warmest year in probably the warmest decade for a thousand years, and was an innocent remark about an abstruse technical point in some graphs showing data from tree rings.

But truth was of no consequence. The lies were out there. This was fairly typical coverage, suggesting some kind of global conspiracy.

[CLIMATE SCAM HEADLINE]

And the climate science community was not responding effectively. I wrote a piece for the online environment magazine run from Yale University, Yale360, headed "Climategate: Anatomy of a Public Relations Disaster".

But the e-mails did raise some real questions, especially about whether the self-policing systems of peer review should be reformed; and about whether in an internet age, it makes sense to keep the world of science so closed.

I cannot go into a forensic examination of the e-mails here. My book discusses the e-mails and their contexts in much greater detail. But what are we to make of this?

In March 2004, Jones wrote to a colleague saying that as a peer reviewer he had "recently rejected two papers from people saying CRU has it wrong over Siberia. Went to town in both reviews, hopefully successfully. If either appears I will be very surprised."

Siberia is a part of the world with very few weather stations, but those stations suggest the most extreme warming of any land area. That was what Jones said. But his critics said the data were skewed by urbanisation near the weather stations. The debate matters. But at least one of those papers about which Jones "went to town" was never published.

The issue here is not who was right, but whether Jones abused a conflict of interest to reject papers questioning his own research.

It was not an isolated example. And in other e-mails, Jones and his colleagues are revealed discussing keeping papers hostile to their own work out of the IPCC reports, which are supposed to be a dispassionate assessment of climate science.

A second issue revealed by the e-mails concerned would be attempts to avoid laws that guarantee freedom of information. In Britain, as in some other countries, freedom of information law is a problem for scientists because it does not distinguish between the duty to share data with other scientists and the wider public. In theory, anyone can ask for and expect to get your data. To my mind, the science community badly failed to see this issue coming when the laws were being drafted. And people like Phil Jones are living with the consequences of that failure. But even so, the law is the law.

As early as 2005, Jones was saying that foreign critics like Steve McIntyre and Ross McKitrick wanted his data and "if they ever hear there is a Freedom of Information (FoI) Act now in the UK, I think I will delete the file rather than send to anyone."

That, incidentally, would have been against the law. But he adopted another approach. By 2007, Jones was writing, "Think I have managed to persuade UEA [the University of East Anglia] to ignore all further FoI requests if the people have anything to do with Climate Audit." Climate Audit is the web site run by McIntyre. That again looks illegal.

Jones oversaw a culture of turning down FoI requests at CRU. The records show that of 105 requests concerning CRU submitted up to December 2009, the university had, by late January 2010, acceded in full to only 10.

This secrecy really got them into hot water over their work on the IPCC assessment published in 2007. In 2008, a British sceptic asked CRU for all e-mails sent by its tree-ring specialist Keith Briffa concerning the report, for which Briffa was a chapter lead author.

The IPCC has its own rules about open review. It archives its formal review process and all exchanges online. But Holland wanted to see if there were any secret e-mail discussions outside that formal process. There were.

Holland's application to see such e-mails was turned down. But the hacked e-mails later revealed that Briffa had a long undisclosed correspondence about an important but unpublished paper by two American researchers, Eugene Wahl and Caspar Ammann.

This correspondence was embarrassing and seemingly a breach of IPCC rules. Clearly, CRU people wanted to hush it up. In one of the most damaging e-mails, Jones asked Mann, "Can you delete any e-mails you may have had with Keith [Briffa] re AR4 [the IPCC report]? Keith will do likewise. Can you also e-mail Gene [Eugene Wahl] and get him to do the same … we will be getting Caspar [Ammann] to do the same."

This e-mail seems to have persuaded the British Information Commissioner's Office that FoI requests at the university were "not dealt with as they should have been under the legislation".

What should we make of all this? Three British inquiries into Climategate all exonerated the scientists of any blatant lack of integrity, but they did make major recommendations to clean up their act.

Geologist and English lord Ron Oxburgh, in a study of CRU's science, criticised its internationally important work on analysing tree rings for evidence of past temperatures. This work was central to the famous hockey stick graph of temperatures over the past thousand years.

He said, "The potential for misleading results arising from selection bias is very great in this area." Yet the researchers were not documenting how those decisions were made, making their conclusions impossible to verify. And he also highlighted how CRU scientists' writing for the IPCC were party to "oversimplifications that omit serious discussion on uncertainties".

The main investigation into research conduct, by retired Scottish civil servant Sir Muir Russell, found a "consistent pattern of failing to display the proper degree of openness", a practice that, he said, had to end. Key data on matters of public importance — like CRU's assembly of 160 years of global thermometre data — cannot be regarded as private property any longer, he said.

To be fair, none of the reports attacked the researchers overall integrity or scientific judgment. Some deliberately left this out of their brief. And those that did include it were not always very convincing.

For me, the most bizarre inquiry was the one conducted by Penn State University into e-mails by its climate scientist Mike Mann to his CRU colleagues. It described his successful career as a researcher and fund-raiser. And then — in a reworking of the old story about the king who had no clothes — said, "Such success would not have been possible had he not met or exceeded the highest standards of his profession."

The king cannot be naked because he is the king. It also sounds alarmingly like a banker's description of the banking system in early 2007. It is also the worst kind of self-policing.

CHAPTER 54

RESEARCH INTEGRITY'S BURNING FUSE: CLIMATE TRUTH BEFORE CHANGE EXPLODES*

Ann Henderson-Sellers

1 Climate Change Reality

"It's not our job to inject a spurious, mythical balance into an unbalanced reality",

– *Alexander Kirby*

Climate change is: real and accelerating; arises because of abuse of 'free' environmental services; generationally-postponed (affects grandchildren); managed by a convention (United Nations Framework Convention on Climate Change — UNFCCC) uniquely initiated by scientists, not nation states; wealth creating; and media titillating. The debacle of the UNFCCC's Conference of the Parties (COP)-15 in Copenhagen in December 2009 was not caused by challenges to the fact of global warming nor to the need for action but by the inability to reach politically acceptable agreements. Climate policy is now a choice between a bad or a very bad future (Stern, 2006; Garnaut, 2008). Between 2005 and 2007, the World Climate Research Programme (WCRP) assembled the largest ever collection of coherent climate change simulations: when the Intergovernmental Panel on Climate Change (IPCC) Fourth Assessment Report (AR4, IPCC 2007a) Working Group 1 (WG1) component was signed off in January 2007, this dataset had been downloaded by more than 1200 researchers (>337 TB data) and used in more than 250 peer-reviewed papers. Ironically, despite this heroic

*Dedicated to Dr. Stephen H. Schneider February 11, 1945 — July 19, 2010.

data management effort, the IPCC WG1 'news' came from a single zero dimensional energy balance model run on a laptop. In this chapter, I argue that, faced by an almost universal preference for obfuscation and denial (Hamilton, 2010), the research integrity question is not about improved quality assurance or more careful management of climate change's massive data deluge. It is whether research managers and funders clearly differentiate between climate change risk research — which requires a meritocracy — as opposed to the response to this risk — which as a social issue demands democratic (or other) community decision-making. I ask, will research leaders continue to support the truth of anthropogenic climate change and prioritise climate change research into really dangerous outcomes that cannot currently be ruled out with less than a 10% chance?

2 Social Tipping Points in Climate Change: 2007 to 2010

"I'm not putting anything in an e-mail that I don't want to appear on the front page of The Washington Times or Fox News."

– *Paul Ehrlich*

Tipping points and positive feedback are two very well known climate phenomena (e.g. McGuffie and Henderson-Sellers, 2004). In everyday life, positive feedback is usually a good thing — say for example from your boss on your performance (Henderson-Sellers, 2010a) but, in systems theory, positive feedbacks exaggerate any disturbance. Positive feedback magnifies changes and can lead to the crossing of thresholds beyond which a new state is entered from which return is not achievable by removing (or reversing) the original disturbance (Lenton *et al.*, 2008).

Climate change entered a different regime in 2007 with the publication of the IPCC's Fourth Assessment Report and the joint award of the Nobel Peace Prize to Al Gore and the IPCC. People no longer asked "whether" human activities are changing the climate but the more urgent questions: "how fast?" "with what impacts?" and "demanding what responses?" In late 2007, a virtuous cycle reinforced the public's recognition of the need for urgent action to mitigate change. Positive feedback, including in the media, showed climate change to be a risk management problem to be solved by all nations (Bali Roadmap, 2007).

The mass media reversed (from virtuous to vicious) the direction of its positive feedback on anthropogenic climate change late in 2009. This time, the character of public perception of anthropogenic climate change was horribly transformed by media coverage of a laughable error in Working Group Two (WG2) IPCC AR4 and selected contents of the Climatic Research Unit of the UK's University of East Anglia (CRU UEA) email. The press pushed public perception past a social tipping point in December 2009. The weak Copenhagen Accord (2009) and the reduced pressure for climate mitigation legislation felt by world leaders are symptoms of the new social state resulting from crossing this irrevocable social threshold (cf. Leiserowitz *et al.*, 2010). Forty years ago, Albert Crewe said: "It is up to the scientific community to point out where they can help.... government cannot be expected to seek our advice, because they are much more accustomed to solving problems by new legislation...." and, "Perhaps better solutions exist... (but) until we can make ourselves heard.... problems are in danger of being grossly underestimated". In my view, the true integrity question about anthropogenic climate change is whether, and with what priority and pressure, urgently required actions are agreed and taken.

3 Research Requires a Meritocracy, Decisions Demand Democracy

"All that is necessary for the triumph of evil is that good men do nothing".
– Edmund Burke

The rules that govern IPCC (to which they sadly failed to adhere themselves during the AR4) include making policy-relevant but no policy-prescriptive statements. The tension between the IPCC prescription: "When risks cannot be well quantified, it is the job of policy to make decisions.... Scientists must make it clear where our job stops and the job of policy begins" (Solomon, 2007) and the democratic requirement "The ultimate policymaker is the public. Unless the public is provided with unfiltered scientific information that accurately reflects the views of the scientific community, policy-making is likely to suffer" (Hansen, 2006) has not been, and may never be, resolved. However, great care must be exercised to avoid both false positive as well as false negative forecast errors (Schneider and Mastrandrea, 2010).

In the unfolding global tragedy of our planetary commons, painfully depicted in Copenhagen in December 2009, national and international leaders are risking the future of the Earth by policy failure; the mass media highlights ethically bankrupt behaviours but fails to demand alternatives; and the assessment instrument of the UNFCCC, the IPCC, has metamorphosed from a useful policy tool into one that, at best, encourages no action and, at worst, justifies inadequate responses (Henderson-Sellers, 2010b). In December 2007, the Director of the WCRP advised the heads of the two UN agencies that jointly sponsor the IPCC to close it because it had "completed its task of assessing the unique challenge that faces humanity" (e.g. IPCC, 2007b). Her recommendation, for an immediate UN action agenda on greenhouse mitigation and adaptation to climate change, was ignored then although by December 2009 the IPCC may have wished it had been accepted. By early 2010, the defence of climate science was insipid as governments and their agencies strove to distance themselves from the tarnished IPCC (e.g. *Nature*, 2010, Netherlands, 2010).

Whether peer review is now a handicap or a benefit is debatable but it is demonstrably unfair (*New Scientist*, 2010; Russell, 2010). For example Schultz (2010) shows that 79% of all atmospheric science journals and virtually all on climate reject more than 30% of submitted papers. Whether rejection of one third of submitted papers is desirable or inadequate may not matter since rejection rate is not an important predictor of the publication's importance (half-life of papers, as measured by citations). While the principle that research be evaluated by experts is widely agreed (Fuller, 2002), peer-review implementation has changed over time (e.g. blind peer reviews of papers in journals began only in 1950s/60s) and is disputed (only 8% of members of the Scientific Research Society agreed that 'peer review works well as it is' (Chubin and Hackett, 1990)). Worse still, Freedom of Information legislation and other even less appropriate laws (such as on tax malfeasance) are being abused in attempts to access climate data and other records (AAAS, 2010; *Guardian*, 2010).

The flaws inherent in the IPCC increasingly render it unhelpful. The top ten challenges for the IPCC are: its linear structure (first 'science', then 'impacts' and, finally, 'mitigation'); peer review (poorly defined and facing an information avalanche: 1,200 exabytes in 2010); protracted gestation (fifth assessment not due until 2014 even though the fourth was out-of-date in late 2006); mandated incapacity to make policy statements; no-preference display of results (failure to "out" bad models); model intercomparison project paradox (gradual community-wide performance improvement masks fundamental failures such as non-conservation: Henderson-Sellers *et al.*, 2008); cost of participation vying with national and laboratory kudos; consensus requirement manifested as fear of highlighting shortcomings and failures (in models and observations); unknown(able) fatness of the probability distribution function tail; and poor handling of published errors and a few "awful emails". Climate prediction has languished near the top of local 'highs' in predictive skill for years (Green, 2006) and, while there are possible routes to improving the current models, few if any groups seem poised to pursue these perhaps because of the massive burden of merely participating in IPCC assessments.

At the moment when other global sustainability challenges are seeking to establish IPCC-like assessments (e.g. Larigauderi and Mooney, 2010), the role of climate assessment is complete. IPCC's failings are well known among participants (Doherty *et al.*, 2009): "adding complexity to models, when some basic elements are not working right (e.g. the hydrological cycle), is not sound science;" "until and unless major (climate) oscillations can be predicted to the extent that they are predictable, regional climate is not a well defined problem. It may never be. If that is the case, then climate science must say so" (Henderson-Sellers, 2008). Research managers and funders should not continue to seek the ill-achieved assurance of consensus as a reason for action (cf. McKibben, 2010).

4 Integrity Paradox: Policy Prescription or People's Ponzi

"The hottest places in hell are reserved for those who, in a period of moral crisis, maintain their neutrality".

– *John F. Kennedy*

Research funders and managers today face a new and unique challenge: support truth against everyone's preference for denial. The public knows climate change is already occurring and its impacts are bad but allows wrong behaviour to persist, accepting media titillation and subsidised fossil fuel resources (e.g. Obama, 2010). The reality that policy is failing to address is that we must limit all future fossil fuel use to less than the amount already consumed: 0.5 trillion tonnes of carbon (Allen *et al.*, 2009). Questioning research integrity is just the latest manifestation of the public's fear of acting on well-known facts and public failure to move to minimize risks about which there is virtual certainty (Ereaut and Segnit, 2006).

In the second half of the 17th century, Blaise Pascal debated with himself about the existence of God (Rescher, 1985). His conclusion was that, despite recognized uncertainty, his best risk reduction strategy was belief. The same Precautionary Principle was embedded in the UNFCCC in 1992

(Schellnhuber, 2010). Despite these guiding ideas, and since at least 1637 (the Tulip Scandal), societies have behaved in highly aberrant ways in a variety of 'Ponzi' schemes.[1] Climate change, which trades the benefits of cheap fossil fuel-derived energy now against future environmental degradation and energy depletion, is rapidly replacing sub-prime loans as the bursting greed bubble (*Economist*, 2010).

In the space of a couple of months from late November 2009 to mid February 2010 two well-known features of the climate system — positive feedback and passing an irrevocable threshold — transformed climate policy from an action agenda (Bali Roadmap, 2007) to an ineffectual and non-binding statement (Copenhagen Accord, 2009). In this period, both emission of greenhouse gases and their impacts accelerated (Global Carbon Project, 2009). Media coverage of investigations into anthropogenic climate change greatly exacerbated public desire: in 2007 for action and in 2009–10 against action.

Climate change research has been found to have integrity by inquiries in the UK, the Netherlands and the USA (UK Parliament, 2010; Oxburgh, 2010; AAAS, 2010; Netherlands, 2010; Russell, 2010). While democracy must uphold law (e.g. Act on CO_2 advertisements banned in the UK in March 2010), the current pretence that climate change or any other research is democratic (e.g. Royal Society UK and Australian Academy, May 2010) is a fabrication. The right of individuals, bodies and groups to make statements cannot be denied but equally their opinions must not override the truth of our current understanding of science (Anderegg *et al.*, 2010). This was most painfully demonstrated this year when the UK Institute of Physics was forced to publicly withdraw its submission to UK parliamentary climate emails inquiry (Guardian, 2010). The 'teaser' for the Singapore Special Session on 'Integrity in the Climate-Change Debate' claims, "Emails written by climate-change researchers, released into the public domain by hackers, have provoked intense scrutiny in the United Kingdom of the integrity of this crucial field of research. Were the researchers simply indiscrete in what they wrote *or is there evidence of efforts to skew or misrepresent data*?" (my italics). I know the organisers of this conference were aware of the findings of the UK Parliamentary and other enquiries and were invited to modify this

[1]Ponzi schemes are fradulent investment systems rewarding investors from their own funds or subscriptions by subsequent participants e.g. the activities of Bernie Madoff.

introduction. Leaders need to uphold truth: not with weak praise (*Nature*, 2010), not by imposing additional quality assurance procedures (Netherlands, 2010) and certainly not by becoming part of the audience titillation trap (cf. Mooney, 2010).

A research integrity statement that clearly differentiates between climate change risk quantification (which is best undertaken by meritocracy) and the response to the identified risk (which, as a social issue, demands community decision-making) could help. So too would prioritisation of research into really dangerous climate change outcomes that cannot currently be ruled out with less than a 10% chance.

"It's no use saying, we are doing our best. You have got to succeed in doing what is necessary."

– Winston Churchill

References

1. AAAS (2010). Statement of the AAAS board of directors concerning the Virginia attorney general's investigation of Prof. Michael Mann's work while on the faculty of University of Virginia. http://www.aaas.org/news/releases/2010/media/0518board_statement_cuccinelli.pdf.
2. Allen, M. R., Frame, D. J., Huntingford, C., Jones, C. D., Lowe, J. A., Meinshausen, M. and Meinshausen, N. (2009). Warming caused by cumulative carbon emissions towards the trillionth tonne. *Nature* 458, 1163–1166, doi:10.1038/nature08019, http://www.nature.com/nature/journal/v458/n7242/full/nature08019.html.
3. Anderegg, W. R. L., Prall, J. W., Harold, J. and Schneider, S. H. (2010). Expert credibility in climate change. PNAS, doi: 10.1073/pnas.1003187107,www.pnas.org/cgi/doi/10.1073/pnas.1003187107.
4. Bali Roadmap (2007). http://unfccc.int/meetings/cop_13/items/4049.php.
5. Copenhagen Accord (2009). http://unfccc.int/resource/docs/2009/cop15/eng/107.pdf.
6. Chubin, D. E. and Hackett, E. J. (1990). Peerless Science: Peer Review and U.S. Science Policy. Albany: State University of New York Press. N.Y, 0791403092, 267 pp.
7. Crewe, A. (1967 and 2007). Science and the war on.... *Physics Today* 20(10): 25–30.
8. Ereaut, G. and Segnit, N. (2006). Warm Words: How are We Telling the Climate Story and Can We Tell it Better? Institute for Public Policy Research, UK, 32 pp.

9. Doherty, S. J., Bojinski, S., Henderson-Sellers, A., Noone, K., Goodrich, D., Bindoff, N. L., Church, J., Hibbard, K. A., Karl, T. R., Kajfez-Bogataj, L., Lynch, A. H., Mason, P. J., Parker, D. E., Prentice, C., Ramaswamy, V., Saunders, R. W., Simmons, A. J., Stafford Smith, M., Steffen, K., Stocker, T. F., Thorne, P. W., Trenberth, K., Verstraete, M. M., Zwiers, F. W. (2008). Lessons learned from IPCC: Developments needed to understand and predict climate change for adaptation. *Bulletin of American Meteorological Society*, April, 2009, doi: 10.1175/2008BAMS2643.1, http://journals.ametsoc.org/doi/abs/10.1175/2008BAMS2643.1.
10. Economist, 24 March 2010. The clouds of unknowing: The science of climate change, http://www.economist.com/displaystory.cfm?story_id=15719298.
11. Fuller, S. (2002). *Knowledge Management Foundations.* Boston: Butterworth-Heinemann, 279 pp.
12. Garnaut, R. (2008). *Garnaut Climate Change Review.* Port Melbourne, Australia: Cambridge University Press.
13. Global Carbon Project (2009). http://www.globalcarbonproject.org/.
14. Greene, M. T. (2006). Looking for a general for some modern major models. *Endeavour*, 30(2): 55–59.
15. Guardian (2010). www.guardian.co.uk/environment/2010/mar/05/climate-emails-institute-of-physics-submission.
16. Hamilton, C. (2010). *Requiem for a Species: Why We Resist the Truth About climate Change.* Sydney: Allen & Unwin.
17. Hansen, J. (2005). Is there still time to avoid 'dangerous anthropogenic interference' with global climate? A Tribute to Charles David Keeling, presentation at American Geophysical Union, 6 December 2005, San Francisco, CA, USA, 14 pp.
18. Henderson-Sellers, A. (2008). The IPCC report: What the lead authors really think? Talking Point Invited Article, ERL, Environmental Research Letters, http://environmentalresearchweb.org/cws/ article/opinion/35820.
19. Henderson-Sellers, A. (2010a). How seriously are we taking climate change? Monitoring climate change communication (Chapter 2), In Y Yu and A Henderson-Sellers, (eds), *Climate Alert Climate Change Monitoring and Strategy*, Sydney: Sydney University Press, pp. 28–65.
20. Henderson-Sellers, A. (2010b). Climatic change: communication changes over this journal's first 'century'. *Clim. Chng.* 100: 215–227, doi:10.1007/s10584-010-9814-9 http://www.springerlink.com/openurl.asp?genre=article&od=doi:10.1007/s10584-010-9814-9.

21. Henderson-Sellers, A., Irannejad, P. and McGuffie, K. (2008). Future desertification and climate change: The need for land-surface system evaluation improvement. *Global Plant. Chng.* 64: 129–138.
22. IPCC (2007a). Climate Change 2007: The Physical Science Basis. Contribution of Working Group I to the *Fourth Assessment Report of the Intergovernmental Panel on Climate Change*, Solomon, S., Qin, D., Manning, M., Chen, Z., Marquis, M., Avery, K. B., Tignor, M. and Miller, H. L. (eds.), Cambridge, United Kingdom and New York, NY, USA: Cambridge University Press.
23. IPCC (2007b). Climate Change 2007: Impacts, Adaptation and Vulnerability. Contribution of Working Group II to the *Fourth Assessment Report of the Intergovernmental Panel on Climate Change*, Parry, M. L., Canziani, O. F., Palutikof, J. P., van der Linden, P. J. and Hanson, C. E. (eds), Cambridge, United Kingdom and New York, NY, USA: Cambridge University Press.
24. Larigauderi, A. and Mooney, H. A. (2010). The Intergovernmental science-policy Platform on Biodiversity and Ecosystem Services: Moving a step closer to an IPCC-like mechanism for biodiversity. *Current Opinion in Environmental Sustainability*, 2: 1–6, doi: 10.1016/j.cosust.2010.02.006.
25. Lenton, T. M., Held, H., Kriegler, E., Hall, J. W., Lucht, W., Rahmstorf, S. and Schellnhuber, H. J. (2008). Tipping elements in the earth's climate system. *Proc Natl Acad Sci.* 105, 1786–1793.
26. Leiserowitz, A., Maibach, E. and Roser-Renouf, C. (2010). Climate change in the American Mind: Americans' global warming beliefs and attitudes in January 2010, Yale University and George Mason University. New Haven, CT: Yale Project on Climate Change, http://environment.yale.edu/uploads/AmericansGlobalWarmingBeliefs2010.pdf.
27. McGuffie, K. and Henderson-Sellers, A. (2004). *A Climate Modelling Primer*, Third edition. John Wiley.
28. McKibben, W. (2010). *Earth: Making a Life on a Tough New Planet.* New York, NY: Times books, Henry Holt & Company.
29. Mooney, C. (2010). The climate trap: this account of last year's 'climategate scandal' inadvertently plays the sceptics game (review of The Climate Files by Fred Pearce). *New Scientist*, 3 July, p. 42.
30. *Nature*, Editorial, Climate of Fear, 2010, 464(7286).
31. New Scientist (2010). Paper Trail: Inside the Stem Cell Wars. Peter Aldhouse, June 9, http://www.newscientist.com/article/mg20627643.700-paper-trail-inside-the-stem-cell-wars.html.

32. Netherlands Environmental Assessment Agency (2010). Assessing an IPCC assessment: An analysis of statements on projected regional impacts in the 2007 report, www.pbl.nl/en. See also Schiermier (2010), Few fishy facts found in climate report: Dutch investigation supports key warnings from the IPCC's most recent assessment. *Nature*, 5 July, doi:10.1038/466170a.
33. Obama, B. (2010). US Presidential address 15 June 2010, http://www.youtube.com/watch?v=Gh76oepKFc8&feature=related.
34. Oxburgh (2010). http://www.realclimate.org/index.php/archives/2010/04/second-cru-inquiry-reports/.
35. Rescher, N. (1985). *Pascal's Wager: An Essay on Practical Reasoning in Philosophical Theology.* Notre Dame: University of Notre Dame Press.
36. Russell, M. (2010). The Independent Climate Change E-mails Review, (Chair Sir Muir Russell), 7 July 2010, 160 pp, http://www.cce-review.org/pdf/FINAL%20REPORT.pdf.
37. Schneider, S. H. and Mastrandrea, M. D. (2010). Risk, uncertainty and assessing dangerous climate change. In S. H. Schneider, N. Rosencranz, M. D. Mastrandrea and K. Kuntz-Duriseti (eds.), *Climate Change Science and Policy*, Chapter 15. Island Press.
38. Schellnhuber, H. J. (2010). Tragic triumph. *Climatic Change* 100: 229–238, doi:10.1007/s10584-010-9838-1.
39. Schultz, D. M. (2010). Rejection rates for journals publishing in the atmospheric sciences. *Bull. Amer. Meteor. Soc.*, doi:10.1175/2009BAMS2908.1.
40. Solomon, S. (2007). The Physical Science Basis: Contribution of Working Group I, Public Presentation, Royal Society, March 9, 2007.
41. Stern, N. (2006). *The Economics of Climate Change.* Cambridge, UK: Cambridge University Press.
42. UK Parliament (2010). House of Commons Science and Technology Committee, The Disclosure of Climate Data from the Climatic Research Unit at the University of East Anglia, Eighth Report of Session 2009–10, http://www.publications.parliament.uk/pa/cm200910/cmselect/cmsctech/387/387i.pdf.

CHAPTER 55

INTEGRITY IN RESEARCH WITH DUAL-USE POTENTIAL

Lida Anestidou

The session *Integrity in Research with Dual-use Potential* was organised in response to the increased demand on life scientists to consider the biosecurity implications of biosciences research. Since the events of September 11, 2001, the perception of bioterrorism threat has increased substantially as the ease of movement across borders and the fundamental principles of collaboration and data sharing have led to a truly globalised research enterprise. Under the so-called "dual-use dilemma", it is now recognised that legitimate research outcomes, methods and tools may be used for harmful purposes as more and more individuals have access to primary resources, technologies and knowledge. Governments have focused on more efficient research oversight[1] and international conventions recommend educating life scientists on matters of national security and the life sciences[2].

Depending on the intent of the user, in principle any research equipment (e.g. fermenters, centrifuges, freeze-dryers), materials (e.g. seed cultures of pathogens, toxins) or technology and knowledge may be used to positively impact the public health, produce much needed pharmaceuticals or advance agriculture, or conversely create biological weapons to support bioterrorism. According to the US National Research Council's report *Biotechnology*

[1]National Science Advisory Board for Biosecurity (2007). Proposed framework for the oversight of dual-use life sciences research: Strategies for minimising the potential misuse of research information. Available at: http://oba.od.nih.gov/biosecurity/pdf/Framework for transmittal 0807_Sept07.pdf.

[2]Sta. Ana, J. L., Frankel, M. S. and Berger, K. M. (2009). Educating scientists about dual-use. *Science*, 326(5957): 1193.

Research in an Age of Terrorism,[3] there are seven classes of experiments that should be subjected to review prior to being undertaken or published because of the potential for misuse of their outcomes. These experiments of concern are as follows:

- Render a vaccine (human or animal) ineffective;
- Confer resistance to therapeutically useful antibiotics or antiviral agents;
- Enhance the virulence of a pathogen of render a nonpathogen virulent;
- Increase transmissibility of a pathogen;
- Alter the host range of a pathogen;
- Enable the evasion of diagnostic/detection modalities; and
- Enable the weaponisation of a biological agent or toxin.

Since 2005, the Medical Research Council of the UK requires grant applicants to consider any dual-use nature of proposed experiments as part of their submission for funding.[4] In 2010, the European Commission's Ethics Review Unit (which is responsible for the ethical review of funding applications to the various European granting agencies) issued a report on research misconduct and misuse[5] that identified three additional areas of research with potential for misuse that the grantees should address:

- Research that can result in stigmatisation and discrimination of individuals;
- Development and application of surveillance technologies; and
- Research with data mining and profiling technologies.

In the face of these new concerns, questions about the practice of science as we have experienced over the last fifty years are emerging. Are some of the values that we hold dear endangered? Should we, for instance, forgo the quest for truth and not engage in some aspects of research for fear of its outcomes? Is there forbidden knowledge and things we should not know, and should we control or restrict the information we publish? In a survey

[3]National Research Council (2004). *Biotechnology Research in an Age of Terrorism*. Washington DC: National Academies Press.

[4]Medical Research Council (2005). MRC Position Statement on Bioterrorism and Biomedical Research. Available at: http://www.mrc.ac.uk/Utilities/Documentrecord/index.htm?d=MRC002538.

[5]European Commission (2010). A Comprehensive Strategy on how to Minimise Research Misconduct and the Potential Misuse of Research in EU-funded Research.

published in 2005, Kempner and colleagues wrote that interviewed scientists acknowledged the right of society to place limits on "what and how science is done".[6] In 2010, Dias and colleagues examined any (negative) effects of new legal mandates on research with select agents in the United States. The authors noted a striking loss of efficiency and a concomitant two- to five-fold increase in the cost of select agent research.[7] Conversely, others have noted the cost of not doing research with these pathogens, implying that the restriction of research with select agents is a threat to public health and national security.[8] It is, therefore, important to ponder whether the dual-use nature of this and other types of biosciences research may hinder the advancement of science.

In a 2009 survey conducted by the American Association for the Advancement of Science and the US National Research Council,[9] participating scientists said they would support self-governance and responsible conduct of research as methods of adequate oversight of research with dual-use potential. This is the central theme that is addressed by this panel: can elements of research integrity and other instruments that preserve the values of science, such as codes of conduct, provide functional oversight of research with dual-use potential? Of the five panelists in Singapore, four contributed to this paper. I thank them all.

---

[6]Kempner, J., Perlis, C. S. and Merz, J. F. (2005). Forbidden knowledge. *Science*, 307(5711): 854.

[7]Dias, M. B., Reyes-Gonzalez, L., Veloso, F. M. and Casman, E. A. (2010). Effects of the USA Patriot Act and the 2002 Bioterrorism Preparedness Act on select agent research in the United States. Proceedings of the National Academy of Sciences, 107(21): 9556.

[8]Franz, D. R., Ehrlich, S. A., Casadevall, A., Imperiale, M. J. and Keim, P. S. (2009). The "nuclearization" of biology is a threat to health and security. *Biosecurity and Bioterrorism*, 7: 243.

[9]National Research Council (2009). *A Survey of Attitudes and Actions on Dual-Use Research in the Life Sciences.* Washington DC: National Academies Press.

CHAPTER 56

GOVERNANCE OPTIONS FOR DUAL-USE RESEARCH

Gerald L Epstein

"Dual-use" science and technology is that which is developed for fully legitimate civilian or scientific purposes, but that has the potential to be abused for the purpose of committing harm. It is hardly a new concept. Concerns about the dual-use nature of science and technology have been with us ever since human beings first learned to forge metal or to sharpen sticks. However, these concerns are perhaps more immediate in the case of modern biology and biotechnology than they are in almost any other field of endeavour. Since pathogenic organisms are self-replicating, the path from research laboratory to catastrophic outcome can be much more direct than it would be for other means of doing harm, which generally require a massive industrial infrastructure to inflict harm over an equivalent area. Even so, biology is not unique in posing dual-use issues, and governance options suitable for addressing security risks related to biology may find application in other areas of science and technology that also pose security concerns.

Whether or not biology's dual-use potential actually constitutes a serious problem is, fortunately, not known for sure. Biological organisms have rarely been deliberately spread to inflict harm. However, many decades ago, weapons programs in several countries demonstrated the ability of biological weapons to cause illness and death over very wide areas. Over the intervening decades, biological science and biotechnology have grown ever more powerful and ever more accessible around the world. Although no legitimate users currently have reason to master all the steps needed to commit acts of bioterrorism or biological warfare, each one of those steps has some legitimate application for civilian or scientific purposes. No aspects of the process are uniquely military or weapons-related, and all are therefore available to anyone with sufficient interest, resources, and expertise.

Moreover, whereas terrorists of the past were believed to be deterred from committing truly horrific acts, lest they inspire revulsion even on the part of constituencies whose sympathies terrorists were trying to elicit, a new generation of terrorists includes at least some who have professed their intention to kill on a mass scale. How can policymakers be sure that no such individuals will ever seek to take advantage of the indiscriminate power to kill at the level made possible by biological weapons?

Some who grant that the misuse of biology poses a serious, albeit to-date largely hypothetical, security concern, nevertheless argue that fundamental research poses little risk. They argue that terrorists are very unlikely to attempt to exploit cutting-edge research results, that the risk of strangling research through restrictive regulations exceeds the risk of bioterrorism, and/or that it is the responsibility of others (i.e. law enforcement or intelligence communities), rather than the research community, to address the malicious use of biology. Such views are short-sighted. First of all, today's cutting-edge research becomes tomorrow craft skill, which becomes the day after's commodity. Any future governance measures will be far more effective if considered earlier rather than later. Secondly, although it is true that yesterday's terrorists have not been particularly interested in biological weapons, the terrorists of tomorrow might well be. Terrorists tend to use the tools with which they are familiar, and in the future, more and more of them will be familiar with biology. Unfortunately, there is nothing inherent to advanced technical education that renders one immune to the forces that lead others to do harm. And thirdly, while the scientific community certainly does not bear unique responsibility for preventing, mitigating, or responding to the risk that biology will be misapplied to do harm, neither can it absolve itself from all responsibility.

Indeed, many different professional communities, including medical, public health, law enforcement, emergency management, scientific research, national security, counterterrorism, veterinary science, and private industry, all have some roles in dealing with this problem. All these communities share two attributes: they all have responsibilities to which they devote most of their time and attention *other than* minimisation of biological risk and many of them would find themselves interacting with each other only in the context of a biological attack. Therefore, it is essential that each of these communities — including the scientific community — spends some time considering what its role should be in preventing, mitigating, or responding to a biological attack, and each of these communities needs to consider the needs, objectives, languages, and modes of operation of others with whom it may need to interact.

The tools that can be applied to address the risks posed by dual-use research are denoted here as *governance options*. But *governance* does not necessarily mean governmental action or regulation. Instead, it refers to any action that anyone might take — whether voluntarily or mandated, whether individually or collectively — to help minimise risk. In fact, traditional tools such as mandated national regulations are poorly suited to deal with problems such as the risks of dual-use biology, which are technically intensive, rapidly evolving, subjective (in that objective criteria defining the degree of risk posed by some proposed projects, and the degree of risk mitigation offered by some policy responses, are essentially impossible to derive), and inherently international. The traditional regulatory process requires specificity in delineating what is banned and what is permitted; it requires time to draft regulations, solicit public comment, make revisions, and put into force; and it has no ability to extend legal authority internationally without negotiation and implementation of a treaty, an instrument that is typically even less able to address these issues than regulation is.

But the scientific community has other tools at its disposal that are better suited to addressing the risks of dual-use research. These tools are typically "softer" ones that do not have the force of law, but that are better able to influence the behaviour of researchers working with dual-use science and technology. These tools include the enunciation of norms and codes of conduct, which will be discussed further in Chapter 59; oversight mechanisms for proposed research projects in which independent reviewers determine that the scientific value of any given experiment outweighs the possible risks that the research results might be applied maliciously; and education and awareness programs to remind biologists and biotechnologists that not everyone who may become aware of the results of their research may have the best interests of all mankind in mind. All components of the scientific community — including individual researchers, research institutions, funding authorities, professional societies, and scientific publications — have the responsibility to ensure that dual-use risks are considered, and mitigated wherever possible.

Whether or not paying attention to the risks of dual-use research will actually lessen the chances that biology will be used to inflict harm, failure to pay attention to these risks poses serious dangers of its own. If the public comes to believe that the scientific community is not paying appropriate attention to foreseeable danger, the scientific community can lose public trust — and a public that no longer trusts that the research community is acting in the public's best interest will be much less willing to fund research or to tolerate its conduct. Such an outcome may be more serious than a bioterrorist attack.

CHAPTER 57

THE ROLE OF LEADERSHIP AND CULTURE WITHIN THE LABORATORY

David R Franz

Historically, most infectious disease researchers working in biocontainment[1] have considered laboratory *safety* their first priority. In the mid-1990s, after an attempt by an acknowledged biological technician[2] to illicitly obtain a strain of the plague bacteria from a national culture collection in the USA, the government instituted the "Select Agent Rule"[3] which called for laboratory registration and pathogen *security* in laboratories working with certain pathogens. After 9–11 and the "anthrax letter attacks",[4] the US government commenced the registration of scientists and the Department of Army instituted a *Personnel Reliability Program.*[5] The National Science Advisory Board for Biosecurity[6] and several national-level committees have subsequently made recommendations regarding US domestic laboratory management policy, seeking an appropriate balance between security and progress in the life sciences.[7] Additional policy and legislative changes are still pending in late 2010. While scientists and policy makers in several nations wrestle with difficult issues surrounding laboratory security and "the insider threat", a revolution in biotechnology — thus an ever

[1] http://en.wikipedia.org/wiki/Biocontainment
[2] Larry Wayne Harris, 1995.
[3] http://www.selectagents.gov/
[4] http://en.wikipedia.org/wiki/2001_anthrax_attacks
[5] http://www.fas.org/irp/doddir/army/ar50-1.pdf
[6] http://oba.od.nih.gov/biosecurity/about_nsabb.html
[7] http://oba.od.nih.gov/biosecurity/meetings/200905T/NSABB%20Final%20 Report%20on%20PR%205-29-09.pdf

increasing capability to do good or harm with biology — roars around the globe.[8] How do we balance our concerns regarding accidental or intentional misuse of these powerful tools, and still exploit them fully for food and energy security and to the benefit of health globally?

The value of and the necessity for sound laboratory safety is broadly understood and accepted by scientists worldwide. Even the physical security of laboratories in which scientists work with especially dangerous pathogens is understood and accepted widely. However, universal agreement has not yet been reached regarding the value or cost of regulations proposed or implemented to protect society from the "insider threat". Although there are numerous specific actions that have been taken in this regard, most are encompassed by the terms *personnel reliability* and *agent accountability*. Some early efforts have been through mandating background investigations, training in ethics and behaviour, and monitoring of medical records to assure that the scientist will not do harm with biology, or facilitate others who seek to do harm. Another general approach has been an attempt to account for pathogens by volume and/or concentration in laboratories, much as can and is done routinely with hazardous chemicals or nuclear/radioactive materials. Accounting for biological materials, which replicate, is a much more difficult proposition than accounting for the other two classes of agents which can be measured by volume or activity. There have also been calls for "two-person"[9] rules in laboratories working with dangerous pathogens. While scientists and laboratory directors have debated the value and cost of the two-person rule, they have broadly accepted the placement of cameras in laboratories to monitor the activities of scientists. It has become clear that regulations such as these, alone, cannot make us totally safe and, if overdone, may actually delay progress for the good of mankind.

Where the technical capability and the biological agents exist, it is "intent" that must serve as the barrier between good and harm in the laboratory. While attempts to protect society from our scientists by instituting regulatory approaches has been widespread and continues, less attention and effort has been placed on the value of *enlightened leadership* and its positive impact on laboratory culture. Leadership is the foundation of any organisational culture. Enlightened, competent leadership can make an

[8]http://www.biologyistechnology.com/

[9]http://www.ncbi.nlm.nih.gov/bookshelf/br.fcgi?book=nap12774&part=ch5

enormous difference in not only the productivity, but also the behavioural aspects of a laboratory through the application of sound management principles. Speaking generally, a culture that builds trust and transparency will be more productive and safer than one that is overly authoritarian and mistrusts its employees. A leader's vision and the open communication of that vision to the staff is important. A good leader hires the best people and delegates responsibility and authority, gives the staff credit for their successes and supports their activities and needs. A good leader is honest with the staff, takes appropriate risks and accepts responsibility for the organisation's failures. Good leaders focus on quality science, laboratory safety, accountability and transparency. The result is a culture of responsibility and accountability that improves safety and security. A possible analogy to a healthy organisational culture is the *Crew Resource Management* (CRM)[10] now widely used in the aviation industry. CRM emphasises the value of openness and transparency, and seeks to instill a confidence that every member of the crew is empowered to point out a safety concern to any member of the crew, including the captain. The resulting culture puts "all eyes" on the safety issue and has significantly improved aviation safety in many parts of the globe. There may be lessons for our laboratories in this model. The likelihood of an insider event cannot be reduced to zero — by any means — but a culture of responsibility, accountably and openness will greatly reduce that likelihood, and will also firmly facilitate progress in research and development. There will always be a legitimate need for regulatory approaches to improving laboratory safety and security, but it is important that we carefully evaluate both the real and intangible costs of such regulation. Making new and stringent rules is the easy way out, but it may not be providing us as much security as we believe. Being an *enlightened leader* can be hard but rewarding work, and not just anyone is capable of building and nurturing a positive culture in the laboratory, but we know it does pay huge dividends in both productivity and safety. While human nature will never allow us to reduce the risk to zero, this approach will very likely also improve our security in the context of the insider threat.

[10] http://en.wikipedia.org/wiki/Crew_resource_management

CHAPTER 58

DUAL-USE RESEARCH, CODES OF CONDUCT, AND THE NATIONAL SCIENCE ADVISORY BOARD FOR BIOSECURITY

F Daniel Davis

Responsibility is an indispensable concept in thinking about and assessing the morally significant actions of individuals and groups, including scientists and the disciplines to which they belong. In the sciences, several questions arise with any invocation of the concept of responsibility: for example, for *what*, exactly, are scientists responsible? And once the *what* has been determined, *how* is the moral sense — some would argue, the *virtue* — of responsibility to be inculcated and cultivated within and among scientists, especially within and among the young?

These and other questions — which are at play in educational programmes for the responsible conduct of research — have been posed with renewed urgency in the last decade, in the wake of 9/11 and the Amerithrax poisonings, events that heightened concerns about dual-use research, that is, biological research with a legitimate scientific purpose that may be misused to pose a threat to public health and/or national security. In 2004, these concerns spurred the establishment of the National Science Advisory Board for Biosecurity (NSABB), which is charged with providing advice and guidance to the government of the USA on policies for the oversight of dual-use research. Three years later, in 2007, NSABB issued its *Proposed Framework for the Oversight of Dual-Use Research* and, therein, set forth its conclusions regarding the question of how best to raise awareness about the dual-use dilemma among scientists, and how best to promote responsibility for actions responsive to the threats of dual-use research. Within its proposed framework, NSABB has envisioned an important, albeit carefully delineated role for codes of conduct.

In this brief essay, I have two tasks. The first is to describe the role of codes of conduct within NSABB's proposed framework for oversight. The second is to offer a more critical account of that role, which I propose to do in relation to another concept, the concept of moral agency.

1 DUAL-USE RESEARCH, CODES OF CONDUCT AND THE NSABB

Comprised of leading individuals from the life sciences and other fields, the NSABB has a multifaceted mission but its central task, as a federal advisory commission, is to provide the US government with expert advice and counsel on policy. What is the scope and what are the limits of the federal government's role in oversight of dual-use research? In the NSABB's deliberations regarding the oversight of dual-use research, this has been a central question that, of course, engenders similar questions about the other principal actors or agents in the life sciences — about the roles, that is, of individual scientists, of the institutions under whose auspices they usually work, of professional societies, of the scholarly journals that publish research findings, and of the corporations, foundations, and government agencies that fund and support research in the life sciences. In 2007, the NSABB published its *Proposed Framework*, a document articulating a set of principles and a division of responsibility for the oversight of dual use research. Within the NSABB framework, the principle that is first and foremost reads as follows:

> The foundation of oversight of dual-use research includes investigator awareness, peer review, and local institutional responsibility. Such oversight allows input directly from the investigators, facilitates timely review, offers appropriate opportunities for public input, and demonstrates to the public that scientists are taking responsibility for their research. (NSABB, 2007, p. 7–8)

Moreover, with regard to the division of responsibility for dual-use research oversight, the Board asserts:

> Researchers are the most critical element in the oversight of dual-use life sciences research... [They] thus have a professional responsibility to be aware of dual-use research issues and concerns, to be aware of the implications of their work and the

> various ways in which information from their work could be misused, and to take steps to minimise misuse of their work. (NSABB, 2007, p. 11)

Thus, within the NSABB oversight framework, scientists themselves play the pivotal role: the ability to prevent the malignant use of research in the life sciences rests largely with the individuals who conduct it. How, then, is awareness of the dual-use dilemma — and responsibility for dual-use research — to be promoted and fostered among scientists themselves?

The NSABB proposes a set of strategies in this respect, focused on outreach and communication and on initiatives at key junctures in the process of conducting and disseminating the findings of dual-use research. Among these strategies are those that focus on the development and promulgation of codes of conduct. Rather than offer a fully articulated model code of its own, however, the NSABB offers a set of considerations for interested individuals, institutions, and organisations; it offers, that is, the raw material essential — but also adaptable — to any particular code of conduct. (See the box below for NSABB's description of the core responsibilities for life scientists conduct dual-use research of concern.)

From Considerations: Core Responsibilities of Life Scientists:

Individuals involved in any stage of life sciences research have an ethical obligation to avoid or minimise the risks and harm that could result from malevolent use of research result. Toward that end, scientists should:

- Assess their own research efforts for dual-use potential and report as appropriate;
- Seek to stay informed of literature, guidance and requirements related to dual-use research;
- Train others to identify dual-use research of concern, manage it appropriately, and communicate it responsibly;
- Serve as role models of responsible behaviour, especially when involved in research that meets the criteria for dual-use research of concern; and
- Be alert to potential misuse of research.

The NSABB is now engaged in several initiatives related to codes of conduct in the life sciences. A key effort is to survey professional societies and institutions to identify barriers to, and proven strategies for the adoption of codes of conduct for dual-use research. Undergirding these initiatives is a conviction about the efficacy of codes of conduct in fostering responsibility for dual-use research — the conviction that codes should be the outgrowth of a "bottom-up", voluntary collective effort by scientists themselves. Such a conviction may be seen as rooted in a particular understanding of moral agency.

2 MORAL AGENCY

Moral agency is the demonstrated capacity to act in ways that warrant the approbation reflected in such words as "good", "right", or "just." As a human capacity, moral agency is complex and multidimensional. It consists in what we often call "character"; but also in moral sensibility, the ability to recognise the salient features of any given situation; in reasoning and practical judgment; in knowledge relevant to the situation and its demands; and in the drive to be both responsible and accountable for one's actions. It is also sensitive — even vulnerable — to the environment and to the demands of the particular situation. Indeed, moral agency is organic in nature: like plant life, if it is to flourish and grow in particular circumstances, it depends upon good roots, in the form of morally formative experiences from the beginnings of life; it depends as well upon soil with the necessary nutrients, that is, upon an environment that lends itself to the cultivation of moral agency. The environment of concern with dual-use is that of the everyday pursuits of new knowledge and new technologies in the life sciences. In this environment and context, moral agency:

- Thrives through a sustained, inward drive on the part of the individual scientist to do the right thing, to be responsible;
- Depends upon thoughtful, morally formative mentoring;
- Requires knowledge and awareness, especially of the salient features of the research itself and of the context in which it is being conducted; and
- Is strengthened by the mutual respect and accountability that ideally defines the relationship between an individual scientist and his/her peers and the discipline (or profession) at large.

Viewed from the perspective of this concept of moral agency, codes of conduct have an important but limited role to play in fostering responsibility for dual-use research within the life sciences. Codes, as such, can do little to compensate for deficiencies of character on the part of any individual; nor can codes alone replace the morally formative process of good mentoring of the young by the experienced. In cultivating the moral agency of individual scientists or groups of scientists, there is no single "magic bullet". There is only an array of interdependent tools, each of which is only as effective as the maker and the wielder of that tool.

Codes of conduct, however, can be utilised as tools for achieving two goals fundamental to moral agency as a capacity of individuals and groups in the life sciences. One goal is to disseminate knowledge and awareness of the dual-use dilemma as elements of the responsible conduct of research. Surveys have shown that there is, in fact, a lack of awareness of the dual-use dilemma among life scientists; grassroots efforts to consider and perhaps adopt codes of conduct are ways, among others, to close this gap in awareness. Another related goal is that of using such efforts, not only to promote awareness but also to encourage a heightened sense of personal and collective responsibility and accountability for the dual-use potential of any given research endeavour. When individuals in any setting decide among themselves to conduct themselves in accordance with a set of agreed upon principles or precepts, the results can be quite positive: shared, mutually reinforced fidelity to such principles and precepts is often far more effective than enforced compliance from above.

Thus, codes of conduct in the life sciences have a limited but nonetheless potentially significant role to play in enhancing knowledge and awareness of the dual-use dilemma on the part of individuals and groups. As with any endeavour to shape moral agency as a human capacity, initiatives to develop and promote codes of conduct demand thoughtful, careful strategies, exquisitely attuned to the specific context. Otherwise, the growth of moral agency — of responsibility and accountability for dual-use research — may be frustrated rather than advanced.

References

1. The National Science Advisory Board for Biosecurity (2007). *Proposed Framework for the Oversight of Dual-Use Life Sciences Research: Strategies for Minimising the Potential Misuse of Research Information.* Washington, DC: NSABB. Available at: http://oba.od.nih.gov/biosecurity/pdf/Framework%20for%20transmittal%200807_Sept07.pdf.

CHAPTER 59

RESEARCH WITH DUAL-USE POTENTIAL IN RCR EDUCATION: IS THERE A ROLE FOR CODES?

Elizabeth Heitman

As national and international policy on research with dual-use potential becomes more extensive and more complex, students in the life sciences increasingly need to learn about the subject early in their careers in order to understand how to identify and minimise the dual-use potential of their own work. The broad scope and multidisciplinary approach of responsible conduct of research (RCR) education seems to make it a particularly appealing context in which to provide trainees with ethical and practical instruction on research with dual-use potential, particularly at the graduate and postgraduate levels.

Current governmental and professional enthusiasm for codes of conduct that addressing research with dual-use potential suggests that RCR education on dual-use issues should include a significant role for codes. Such a role, and its definition, depends on the answers to some preliminary questions:

- Is RCR education indeed the proper venue for teaching about dual-use issues? If so, what goals should RCR education serve with regard to dual-use issues?
- What is the appropriate audience with which to address dual-use issues in RCR education, and in what settings should such instruction be provided?
- How are codes used in RCR education generally?
- What appropriate codes exist?

In many respects, RCR education offers a logical context for instruction about dual-use issues because it is intended to develop young researchers' knowledge and understanding of the professional norms of research, and raise their awareness of and sense of responsibility for the broader scientific and social effects of their more immediate work.

RCR educators have long considered laboratory safety, biosafety, environmental safety, and the social role of science to be important areas for instruction (Heitman and Bulger, 2005). Each of these domains involves dual-use issues, and each is now explicitly included in the National Institutes of Health's policy as a core areas for RCR instruction (NIH, 2009). In its most recent biosecurity workshop, the Inter-Academy Panel (IAP) on International Issues, comprising over 100 national academies of science, included RCR education as a key platform through which to promote education on dual-use issues for life sciences (NRC, 2010).

As a case study in the complexity of contemporary research in the life sciences, consideration of dual-use issues can also illustrate important issues in research integrity, such as the interrelatedness of ethical and practical issues in science, the uncertain impact of scientific discovery and development, and how professional self-governance can effectively address complex questions beyond the reach of law.

The goals of instruction on research with dual-use potential are not fully articulated, whether in the context of RCR education or elsewhere. In many settings, efforts to establish instruction programs are driven by security concerns as a way to alert trainees to possible malevolent uses of basic and applied life sciences research, particularly bioterrorism (NCR, 2010; NSABB, 2008). The US National Science Advisory Board for Biosecurity (NSABB) and IAP expect that increased awareness of the potential nefarious uses of some research will allow trainees to reshape or redirect their work, or to ask for assistance in doing so, as well as maximise their participation in effective oversight. Professional organisations like the American Society for Microbiology also anticipate that education on dual-use issues will allow trainees to participate more fully in promoting their professional responsibilities to society generally (ASM, 2005b).

Typically, the anticipated audience for RCR education on dual-use issues is trainees in the "life sciences" but the use of that term may obscure the many different disciplines and specialties to which it might apply, from microbiology to genetics to agricultural engineering. Each discipline has its own professional identify and possible concerns about research with dual-use potential. At the introductory level, it may be particularly important to identify how research in all scientific fields has the potential for harmful as well as beneficial application. While instruction on and discussion of national and international policy is largely theoretical and classroom based, some dual-use issues, such as biosafety skills and data management techniques, may be best considered in the lab as they arise in practice.

The use of codes in RCR education is likely widespread, but actual practice is unknown. The study of codes can provide an introduction to a profession's self-definition, ethical standards, and process of self-regulation for new trainees. Such professional codes of ethics as the Hippocratic Oath, the American Society for Microbiology's (ASM) *Code of Ethics* (ASM, 2005a), and the American Statistical Association's *Guidelines for Statistical Practice* (ASA, 1999) articulate the respective profession's values and history. Codes typically proclaim the profession's dedication to public welfare and others' benefit, as well as the need to protect the profession and its work from others who might co-opt their skills or use them improperly.

RCR educators can point out contemporary ethical challenges faced by specific fields by having trainees analyse multiple editions of codes over time or compare codes across disciplines. Students can be assigned to draft a code for their class or discipline as a way to introduce key professional values and the importance of professional self-regulation. Codes can serve as important ethical guides in discussion of case studies, a common feature of RCR education in the USA (Heitman, 2002). Using codes to teach about dual-use issues can help trainees recognise ethical grey areas in their work, as well as the need for judgment and moral imagination in interpreting professional codes in specific circumstances.

A variety of codes and calls for codes have been developed in the past decade that can be used in teaching about dual-use issues. Sommerville and Atlas' "Code of Ethics for the Life Sciences" (Sommerville and Atlas, 2005) provides a useful overview to the role of codes in professional ethics, and ASM's natural concern for the dual-use potential of is members' work makes its 2005 code a key document for related RCR education (ASM, 2005a). The IAP Biosecurity Working Group's 2005 guidance on codes (IAP, 2005), together with its earlier documents on the content, development, and adoption of codes and their use in education (IAP, 2008) provide an international perspective on dual-use issues, whereas the NSABB's Guidelines for Codes of Ethics (NSABB, 2008) offers insight into the educational role of developing a code on dual-use issues, rather than a code itself.

The way in which codes that address dual-use issues are presented also affects trainees' acceptance and adoption of relevant standards of practice. Studies on the effect of academic honour codes and computer-use codes at major universities across the USA suggest that codes presented simply as enforceable standards have limited effect on students' behaviour (Synfax Reports, 2000; 2002). To foster understanding of the purpose and goals behind a professional code, it is essential to encourage its discussion in multiple contexts and to demonstrate its clear institutional support.

The use of professional codes to address ethical questions and policy issues about dual-use research should prompt students to reflect on their own activities and work going on around them. Ideally, it should promote dialogue about standards and common challenges, not only among trainees but among more senior researchers. However, after studying codes related to research ethics, trainees may be quick to note inconsistent practices around them. Without ongoing institutional attention to the standards of practice in their own labs and the broader research environment, trainees may become cynical about codes and professional standards in "real life", and see only others' misbehaviour that exploits apparent loopholes.

RCR educators should use and develop curricular material that is relevant to the specific disciplines in which they teach, as well as material that illustrates the common practices and purposes of the broader life sciences. One such comprehensive training module on ethical and policy issues in research with dual-use potential is available from the Policy, Ethics and Law (PEL) Core of the Southeast Regional Center of Excellence in Emerging Infections and Biodefense (SERCEB). SERCEB PEL's online tutorial "The Dual-Use Dilemma in Biological Research" (www.serceb.org/dualuse.htm), is designed to introduce research trainees, as well as their faculty advisors, program directors, and senior colleagues, to the scientific, ethical, and policy questions that they may need to consider in designing, conducting, and publishing their work. The module's attention to codes as well as more formal regulation offers an excellent opportunity for educators to discuss the role of professional codes and professional self-governance in an increasingly regulated research environment.

References

1. American Society for Microbiology (2005a). *Code of Ethics.* Available at: http://www.asm.org/ccLibraryFiles/FILENAME/000000001596/ASMCodeofEthics05.pdf.
2. American Society for Microbiology (2005b). ASM Presentation at the Biological Weapons Convention Meeting, Geneva, 16 June 2005. Available at: http://www.asm.org/index.php/policy/june-16-2005-

asm-presentation-at-the-biological-weapons-convention-meeting-geneva.html.

3. American Statistical Association (1999). *Ethical Guidelines for Statistical Practice*. Available at: http://www.amstat.org/about/ethicalguidelines.cfm.
4. Heitman, E. (2002). Using cases in the study of ethics. In Bulger, R. E., Heitman, E., Reiser, S. J. *The Ethical Dimensions of the Biological and Health Sciences*, 2nd ed. New York: Cambridge University Press. pp. 349–363.
5. Heitman, E. and Bulger, R. E. (2005). Assessing the educational literature in the responsible conduct of research for core content. *Accountability in Research*, 12: 207–224.
6. Interacademy Panel on International Issues (2005). IAP statement on Biosecurity. Available at: http://www.interacademies.net/File.aspx?id=5401.
7. Interacademy Panel on International Issues (2008). Statements, 1993–2008. Available at http://www.interacademies.net/File.aspx?id=8487.
8. National Institutes of Health (2009). Update on the Requirement for Instruction in the Responsible Conduct of Research. NOT-OD-10-019, 24 November 2009.
9. National Research Council (2010). *Challenges and Opportunities for Education About Dual-Use Issues in the Life Sciences*. Washington, DC: National Academies Press.
10. National Science Advisory Board for Biosecurity (2008). Strategic plan for outreach and education on dual-use research issues. 10 December 2008.
11. Sommerville, M. A. and Atlas, R. M. (2005). A code of ethics for the life sciences. *Science*, 307: 1881–1882.
12. Synfax Weekly Report (2000). New research on academic integrity: The success of "modified" honor codes. *Synfax Weekly Report*, 0–17: 975. Available at: http://collegepubs.com/ethics?PHPSESSID=91f45332b6acd8c3817006a9a7b91611#2000.
13. Synfax Weekly Report (2002). Model code of student conduct online. *Synfax Weekly Report*, 2–19: 3007. Available at: http://collegepubs.com/ethics?PHPSESSID=91f45332b6acd8c3817006a9a7b91611#2002.

APPENDIX
SINGAPORE STATEMENT ON RESEARCH INTEGRITY

Preamble. The value and benefits of research are vitally dependent on the integrity of research. While there can be and are national and disciplinary differences in the way research is organised and conducted, there are also principles and professional responsibilities that are fundamental to the integrity of research wherever it is undertaken.

1 Principles

— *Honesty* in all aspects of research
— *Accountability* in the conduct of research
— *Professional courtesy and fairness* in working with others
— *Good stewardship* of research on behalf of others

2 Responsibilities

1. ***Integrity:*** Researchers should take responsibility for the trustworthiness of their research.

2. ***Adherence to Regulations:*** Researchers should be aware of and adhere to regulations and policies related to research.

3. ***Research Methods:*** Researchers should employ appropriate research methods, base conclusions on critical analysis of the evidence and report findings and interpretations fully and objectively.

4. ***Research Records:*** Researchers should keep clear, accurate records of all research in ways that will allow verification and replication of their work by others.

The Singapore Statement on Research Integrity was developed as part of the 2nd World Conference on Research Integrity, 21–24 July 2010, in Singapore, as a global guide to the responsible conduct of research. It is not a regulatory document and does not represent the official policies of the countries and organisations that funded and/or participated in the Conference. For official policies, guidance, and regulations relating to research integrity, appropriate national bodies and organisations should be consulted. Available at: www.singaporestatement.org.

5. ***Research Findings:*** Researchers should share data and findings openly and promptly, as soon as they have had an opportunity to establish priority and ownership claims.

6. ***Authorship:*** Researchers should take responsibility for their contributions to all publications, funding applications, reports and other representations of their research. Lists of authors should include all those and only those who meet applicable authorship criteria.

7. ***Publication Acknowledgement:*** Researchers should acknowledge in publications the names and roles of those who made significant contributions to the research, including writers, funders, sponsors, and others, but do not meet authorship criteria.

8. ***Peer Review:*** Researchers should provide fair, prompt and rigorous evaluations and respect confidentiality when reviewing others' work.

9. ***Conflict of Interest:*** Researchers should disclose financial and other conflicts of interest that could compromise the trustworthiness of their work in research proposals, publications and public communications as well as in all review activities.

10. ***Public Communication:*** Researchers should limit professional comments to their recognised expertise when engaged in public discussions about the application and importance of research findings and clearly distinguish professional comments from opinions based on personal views.

11. ***Reporting Irresponsible Research Practices:*** Researchers should report to the appropriate authorities any suspected research misconduct, including fabrication, falsification or plagiarism, and other irresponsible research practices that undermine the trustworthiness of research, such as carelessness, improperly listing authors, failing to report conflicting data, or the use of misleading analytical methods.

12. ***Responding to Irresponsible Research Practices:*** Research institutions, as well as journals, professional organisations and agencies that have commitments to research, should have procedures for responding to allegations of misconduct and other irresponsible research practices and for protecting those who report such behaviour in good faith. When misconduct or other irresponsible research practice is confirmed, appropriate actions should be taken promptly, including correcting the research record.

13. ***Research Environments:*** Research institutions should create and sustain environments that encourage integrity through education, clear

policies, and reasonable standards for advancement, while fostering work environments that support research integrity.

14. ***Societal Considerations:*** Researchers and research institutions should recognise that they have an ethical obligation to weigh societal benefits against risks inherent in their work.

LIST OF CONTRIBUTORS

Ng Eng Hen
Former Minister for Education and currently Minister for Defence
Government of Singapore
Singapore 669645

Su Guaning
Former President and currently President Emeritus
Nanyang Technological University
Singapore 639798

Lim Chuan Poh
Chairman, Agency for Science
Technology and Research
Singapore 138632

Seeram Ramakrishna
National University of Singapore
Singapore 119077

Howard Hunter
Former President and currently Professor of Law
Singapore Management University
Singapore 188065

Christine C. Boesz
Adviser, U.S. Government Accountability Expert
Washington
United States DC 20548

Howard Alper
Visiting Executive at the International Development Research Centre (IDRC) and Distinguished University Professor at the University of Ottawa
Ottawa
Canada K1N 6N5

Lee Eng Hin
Executive Director of the Biomedical Research Council
Agency for Science
Technology and Research
Singapore 138668

Ian Halliday
President, European Science Foundation and Chief Executive of SUPA (The Scottish Universities Physics Alliance)
Stroud
United Kingdom EH9 3JZ

Jean-Pierre Alix
Adviser to Chairman, Science in Society Programme
Centre National de la Recherche Scientifique (CNRS)
Paris
France 39106

Ronald Heslegrave
Senior Scientist, Research Ethics
University Health Network
University of Toronto, Toronto
Canada M5G 1Z5

Sylvia Rumball
Professor Emeritus (Research Ethics)
Research Ethics Office, Massey University
Palmerston North
New Zealand 4442

John O'Neill
Director of Research Ethics and Professor of Teacher Education
Massey University
Palmerston North
New Zealand 4442

Emilio Bossi
President of the Scientific Integrity Committee of the
Swiss Academies of Sciences
Berne
Switzerland

Tohru Masui
Director, Department of Disease Bioresources Research
Research Leader, Laboratory of Rare Disease
Biospecimen and Office of Policy and Ethics Research
National Institute of Biomedical Innovation, Osaka
Ibaraki-shi
Japan 567-0085

Ren Yi
Director, Higher Degree Research Office
Macquarie University
Sydney
Australia 4350

Eero Vuorio
Director, Biocenter Finland
Helsinki
Finland FI-00014

Dirk G. de Hen
Chair, European Network of Research Integrity Offices
Secretary, National Board for Science Integrity
Amsterdam
Netherlands

Boris Yudin
Head of Department, Institute of Philosophy
Russian Academy of Sciences
Moscow
Russian Federation 119992

Daniele Fanelli
Leverhulme Fellow and former
Marie-Curie Intra-European Fellow
University of Edinburgh
Edinburgh
United Kingdom EH1 1LZ

David Vaux
Director, La Trobe Institute of Molecular Science
Melbourne
Australia VIC 3086

Ben Martin
Specialist Adviser, House of Lords Science
and Technology Committee and Editor
Research Policy, University of Sussex
Brighton
United Kingdom BN1 9QE

Edvard Kruglyakov
Counsellor, Russian Academy of Science
Budker Institute of Nuclear Physics, Russia
Novosibirsk
Russian Federation

José Cuellar
National Autonomous University of Mexico
President, Mexican College of Physicians
of the National Institutes of Health
Mexico City
Mexico

Melissa Anderson
Professor of Higher Education and Affiliate Faculty
Center for Bioethics, University of Minnesota
Minneapolis
United States MN 55455

Marta Shaw
Comparative & International Development Education
University of Minnesota
Minneapolis
United States MN 55455

Matthias Kaiser
Director, National Committee for
Research Ethics in Science and Technology
Oslo
Norway NO-0105

John Sulston
Chair, Institute of Science Ethics and Innovation
University of Manchester
Manchester
United Kingdom M13 9PL

Pieter Drenth
Honorary President, All European Academies (ALLEA)
Amsterdam
Netherlands 1011 JV

Frank Wells
Ethics Officer & Chairman
European Forum for Good Clinical Practice Ethics Working Party
Ipswich
United Kingdom IP9 2JJ

Ragnvald Kalleberg
Professor of Sociology, University of Oslo
Oslo
Norway 316

Ashima Anand
Principal Investigator
Exertional Breathlessness Studies Laboratory (DST)
V.P. Chest Institute, Delhi University
Delhi
India 31073 Cedex 7

Ping Sun
Program Coordinator, Office of Research Integrity
Ministry of Science and Technology, China
Beijing
China 100862

Timothy Dyke
Executive Director, Quality and Regulation Branch
National Health and Medical Research Council (NHMRC)
Canberra
Australia ACT 2601

Michael Kalichman
Director, Research Ethics Program
University of California, San Diego
La Jolla
United States CA 92093

Philip Langlais
Vice Provost, Graduate Studies & Research
Old Dominion University
Norfolk
United States VA 23529

Gerlinde Sponholz
Ethicist in Medicine and Research
Representative of the German Ombudsman for Science
(Ombudsman fuer die Wissenschaft)
Germany

Sonia Vasconcelos
Postdoctoral Fellow, Federal University of Rio de Janeiro, Brazil
Rio de Janeiro
Brazil 21941 590

Daniel Vasgird
Director, Office of Research Integrity and Compliance
West Virginia University
Morgantown
United States 26506

Bruce McKellar
Professor of Theoretical Physics at Melbourne University & Chair
International Council of Science Regional
Committee for Asia and the Pacific
Melbourne
AustraliaVIC 3010

Tetsuji Iseda
Associate Professor
Department of Philosophy and History of Science
Kyoto University
Kyoto
Japan 606-8501

Iekuni Ichikawa
Professor, Department of Bioethics
Tokai University School of Medicine
Hiratsuka
Japan 259-1193

Masaru Motojima
Department of Bioethics, Tokai University School of Medicine
Hiratsuka
Japan 259-1193

Makoto Asashima
Director of Organ Development Research
Laboratory National Institute of Advanced
Industrial Science and Technology (AIST)
Ibaraki
Japan 300072

Nils Axelsen
Ombudsman for Research Integrity
at Statens Serum Institut in Copenhagen
Copenhagen
Denmark 12227-010

Simon Bain
Australian National University
Canberra
Australia ACT 0200

Mary Ritter
Chief Executive Officer, EIT Knowledge and Innovation Centre —
Climate and former Pro-Rector, International Affairs
Imperial College London
London
United Kingdom SW7 2AZ

Stephen Webster
Senior Lecturer, Imperial College London
London
United Kingdom SW7 2AZ

Nicholas Steneck
Director, Research Ethics and Integrity Program
Michigan Institute for Clinical and Health
Research & Professor Emeritus of History
University of Michigan
Ann Arbor
United States MI 48109

Sabine Kleinert
Senior Executive Editor
The Lancet & Vice Chair
Committee on Publication Ethics (COPE)
London
United Kingdom NW1 7BY

Elizabeth Wager
Chair, Committee on Publication Ethics (COPE)
Princes Risborough
United Kingdom HP27 9DE

Iveta Simera
Senior Project Manager, EQUATOR Network
Oxford
United Kingdom OX2 6UD

David Moher
Senior Scientist, Ottawa Methods Centre
Clinical Epidemiology Program
Ottawa Health Research Institute (OHRI) & Associate Professor
Department of Epidemiology and Community Medicine
University of Ottawa
Ottawa
Canada K1N 6N5

Douglas Arnold
McKnight Presidential Professor of Mathematics
University of Minnesota & President
Society for Industrial and Applied Mathematics (SIAM)
Minneapolis
United States MN 55455

Vladimir Vlassov
Professor of Moscow Medical Academy
Moscow
Russian Federation 101000

Mark Frankel
Director of the Scientific Freedom
Responsibility and Law Program
American Association for the Advancement of science (AAAS)
Washington
United States DC 20005

Fred Pearce
Freelance Author and Journalist
London
United Kingdom

Ann Henderson-Sellers
Professor & ARC Research Fellow
Department of Environment and Geography
Macquarie University
Sydney
Australia NSW 2109

Lida Anestidou
Senior Program Officer, Institute for Laboratory Animal Research
National Academy of Sciences, USA
Washington
United States DC 20001

Gerald L. Epstein
Director, Center for Science, Technology
and Security Policy
American Association for the Advancement of Science
Washington
United States DC 20005

Daniel Davis
Senior Bioethics Policy Advisor
Office of Biotechnology Activities/Office of the Director
National Institutes of Health
Maryland
United States 20892